PHYSICS AND MAN

A Note from the Publisher

This volume was printed directly from a typescript prepared by the author, who takes full responsibility for its content and appearance. The Publisher has not performed his usual functions of reviewing, editing, typesetting, and proofreading the material prior to publication.

The Publisher fully endorses this informal and quick method of publishing lecture notes at a moderate price, and he wishes to thank the author for preparing the material for publication.

PHYSICS AND MAN

ROBERT KARPLUS

University of California, Berkeley

W. A. BENJAMIN, INC.
New York   1970

PHYSICS AND MAN

Copyright © 1970 by W. A. Benjamin, Inc.

Standard Book Number 8053–5213–9 (Paperback Edition)
Library of Congress Catalog Card Number 77–12146
Manufactured in the United States of America
12345K3210

W. A. BENJAMIN, INC.
New York, New York 10016

Preface

What is physics? The answer to this question as provided by a textbook may seem formal and remote from human concerns, since physical models and theories are abstractions from experience. In this small volume of selected readings I have tried to give a different kind of answer, an answer that reveals some of the personal, social, and humanistic elements in physics. Accordingly, the articles stress interdisciplinary matters and do not expound physical principles or physical phenomena in technical terms.

The theme of the collection is established by Alan Holden's article "Artistic Invitations to the Study of Physics." Holden's ideas may surprise you, for his message is that you should not take physics too seriously or too purely. And that is very good advice to physics students and to physics teachers!

The selections are grouped into seven chapters, each introduced by a brief statement identifying the common significance of the items; yet, the articles are complete in themselves. You may therefore look for selections that supplement current material in your textbook; references to my textbook *Introductory Physics, A Model Approach* (W. A. Benjamin, New York, 1969) are included in several chapter introductions. You may instead prefer to read the book systematically from the beginning, or you may browse to follow your mood or interest of the moment. Only in the chapter on nuclear energy have I made an effort to follow the historical developments of the past thirty years.

I hope you will make an effort to become acquainted with the forty-nine individuals represented in this collection. About half of them are natural scientists who have added an outside interest to their professional work and describe this interest rather than their specialty. The remainder are non-scientists whose lives are touched by physics and who can therefore illuminate the interaction of physics with other human concerns.

Finding and selecting the articles for this book gave me a great deal of personal pleasure. First of all, I was compelled to read much interesting material I would otherwise have never encountered. Second, I discovered unsuspected interests and talents among my friends, many of whom have written about interdisciplinary problems in addition to the work with which I was familiar. Third, I corresponded and became acquainted with some of the contributors I had not known before. And, fourth, I became aware of a growing group of individuals, deeply concerned about the role of science in our world, with whom I felt a community of interest. I am indebted to all the contributors and their publishers for allowing their work to be used.

ROBERT KARPLUS

Berkeley, California
February, 1970

Contents

(all page numbers are text numbers shown at bottom outside each page. Because all articles are reprints, original page numbers are also shown.)

Introduction

About the Contributor

ALAN N. HOLDEN (1904-   ) has served on the technical staff of the Bell
Telephone Laboratories for more than thirty years. His life-long study of
crystal growth and a recent interest in physics teaching led him to write a
book for laymen (Crystals and Crystal Growing, with coauthor Phylis Singer,
Anchor Books of Doubleday and Co., Inc., Garden City, New York, 1960)
and to produce an educational film on the structure of matter with the
Physical Sciences Study Committee.

Reprinted from AMERICAN JOURNAL OF PHYSICS, Vol. 36, No. 12, 1082–1087, December 1968
Printed in U S. A.

Artistic Invitations to the Study of Physics

ALAN HOLDEN

Bell Telephone Laboratories Incorporated, Murray Hill, New Jersey 07974

The Robert A. Millikan Lecture Award is made annually to a physicist chosen by a special committee of the American Association of Physics Teachers because of his creative and imaginative contributions to the teaching of physics. The lecturer selected is one whom the Association wishes to honor at the Summer Meeting in the same way that it honors the Oersted Medalist at the time of the Annual Meeting. The award is made possible through the annual support of Prentice-Hall, Inc.; the first award was made in 1964.

The recipient of this award for 1968 is Alan Holden, Member of the Technical Staff, Bell Telephone Laboratories.

In 1791 when he was a boy of seventeen years, studying at Christ's Hospital, Samuel Taylor Coleridge wrote to his brother, "I have often been surprised that Mathematics, the quintessence of Truth, should have found admirers so few and so languid. Frequent consideration and minute scrutiny have at length unravelled the cause; *viz.* that though Reason is luxuriating in its proper Paradise, Imagination is wearily traveling on a dreary desert. To assist Reason by the stimulus of Imagination is the design of the following production."

The production was a problem of Euclid put in verse. But we can guess that the talented and charming beatnik was exposed later to a course in physics. Dreaming of himself as the Wedding Guest, he wrote, "He went like one that hath been stunned/ And is of sense forlorn./ A sadder and a wiser man/ He rose the morrow morn." Succeeding years have seen that wisdom spread: Don't study physics unless you're a physicist. Today physics acquires many practitioners, but aside from them its admirers are few and languid.

I once thought there was danger in this state of affairs—the danger of constructing a modern counterpart of the medieval priesthood, believed and not understood—but a composer of music divested me of that arrogance. Said he, "Don't kid yourself; people look upon a physicist as they look upon a garage mechanic. If you want a job of physics done, you go to a physicist and pay him, and the little chap does it for you."

In fact, a healthy musical composer looks at his own trade in that way: He is a craftsman for hire. For several years J. S. Bach knocked off a new church cantata almost every week and trained his choir to put it on at the Thomaskirche next Sunday. No shenanigans like opus numbers: About 200 have come to light; an estimated 100 are lost.

Nowadays we, like the imperial Romans, discern barbarians at ever-increasing distances; we hire physicists to thwart them. Would you go to the moon? Given a little time and lots of money, the physicists can get you there. "Science bust mars Lisa opening" read a newspaper headline when the elevators conked out at the unveiling of the celebrated painting in Washington.

The composer's music may give pleasure to many or to few. There is a host of Gershwin lovers, a handful of Schönberg lovers. In any case the composer accepts without complaint[1] that his music will not give pleasure to everybody. But he believes that, if people will only listen, some of them will enjoy his music. And the physicist correspondingly believes that, if people will pay attention, he can shed for some of them a little light on what happens in the world—a physicist's kind of light—indeed his personal kind of light. Which prepares the ground for:

PROPOSITION 1—PHYSICS IS NOT A SINGLE THING

By a robust definition current in earlier times, an axiom is a self-evident truth. The proceedings of the Carleton Conference of 1956 make clear that my first proposition is not self-evident. That was a good conference. It promulgated the idea, among others, that "... these seven principles and concepts outline the minimum content which *any* introductory course must encompass...."

I attended, and subscribed to that notion, and have since come to question it.[2]

---

[1] To be sure Bach once wrote, "My masters are a poor folk, with little music in them," but we all have our low moments.

There is a physics that is a single thing. It is the flesh-and-bloodless body of understanding, visualized by most young students, which descends like manna from heaven and comes to rest in a textbook. But physics, like music, is made by people; and physicists, like composers, come in all kinds; and therefore their physics comes in all kinds, too. Look about you at the practicing physicists in your acquaintance and you know this.

It is hard to believe that physics was the same thing to James Joule and his contemporary, Rudolf Clausius. Professor Andrade tells that, when Willy Wien said to Ernest Rutherford, "But no Anglo-Saxon can understand relativity," Rutherford replied, "No, they have too much sense." J. J. Thomson wrote an Adams Prize essay examining the mathematical properties of toroidal vortices. R. W. Wood, so I have heard, projected such vortices toward his audience and knocked the hats off the ladies. Can physics have been the same to both? When Wolfgang Pauli visited Bell Telephone Laboratories, he examined the ongoing physical researches of which the institution was proudest. Asked what he thought of them he replied, I am told, "Ach, das ist alles triviale": they were irrelevant to the problems that interested him.

If physics is not a single thing, trying to teach it as if it were is mistaken. And if physicists come in all kinds, so (in spades) do physics students. At least we hope they do. Or, better, we *say* we hope they do. Our teaching practices may in fact be limiting them to *one* kind, the kind that we are relatively comfortable teaching.

There is a central core of physics that every physicist must learn, of course.[3] Must Coleridge

---

[2] A good conference tackles large questions, works hard, and has the courage to come up with responsible answers. If a few of the answers turn out to be questionable, that shouldn't discredit the conference. The report of this one can be read in Am. J. Phys. **25,** 417 (1957).

[3] Of course? Well, maybe; but maybe not by taking the courses that teach it. Count Rumford? Michael Faraday? Ludwig von Helmholtz? Perhaps it is crippling not to have passed through these courses; then much of the central core of physics has been constructed by cripples.

learn it too in order to get the hang of the thing? If so, must he learn it in the same way? Answering these two questions is impossible; but examining them is important nevertheless, and some of the examination must be conducted in the light of:

PROPOSITION 2—TEACHING PHYSICS IS A BLACK ART

Says my composer's garage mechanic, "Sure I'll fix your heap; but wouldn't you like to have me tell you about it?" Like as not, the answer is *no*. There is a Principle of Least Knowledge under which we all operate. Greek epigraphy, cryptogamic botany, Provençal philology—if we began learning about them, we might get interested, and that would be fatal. A microcosm of the menace is in the child's book review, "This book told me more about penguins than I want to know." God may not have intended us to be a learned folk: We learn what we need to learn in order to live (which God knows is quite a lot), and we resist the rest.

We resist our resistance, too, however, and try to learn and to teach. It is not clear which effort is the harder. But if you are ever assaulted with the abominable teacher-blasting aphorism, you might fling back the paraphrase, "Those who can, teach; those who can't, learn."

Perhaps my second proposition *is* a self-evident truth: I have found nobody who does not agree that teaching—teaching physics or anything else—is an art.[4] "Black" simply describes the species of art to which teaching belongs. Unfortunately the axiom, once stated, is commonly allowed to sit unused for any deductive purpose.

The "black arts" are the arts of presenting those other arts that live only in performance— drama, ballet, music, and the like. The words of a script, the notes of a score, are directions (necessarily quite incomplete) for performance. The playwright depends on directors and actors, the composer on conductors and instrumentalists, to bring his work to life. Between performances the

---

[4] Learning is an art also, as anybody who has tried to examine and improve his own learning processes will attest.

work sleeps, and each performance is a little different.

Insofar as physics is a body of diverse materials to be taught, the physicists who made it are analogous to the composers, and a physics teacher to a conductor. The performance is the whole coordinated bit—texts, lectures, laboratory, classroom work, office hours, exercises, and tests. Or perhaps computer programs, film loops, teaching machines, and closed circuit TV shows. And the art gives life by performance before one student, or a dozen, or a thousand.

A conductor makes a program of music—of pieces written usually by different composers—and good program making is one facet of his art. At his best he knows the music so well, even if it is new, that he can conduct it without score and so pay better attention to the performance. A physics teacher makes a course out of bits of physics constructed by different physicists, and good course making is . . . pursue the resemblance yourself.

Conductors sometimes arrange pieces of music for performance on other instruments than the composers intended. The results are often shocking—Debussy's "Reflets dans l'Eau" blasted from a pipe organ. I think I would feel a similar shock if somebody tried to teach wave mechanics by experiments with subharmonic oscillators. Here we come upon tastes, the things for which there is no accounting, and mine are Apollonian, not Dionysian. I prefer Bach to Handel, Matisse to van Gogh, David Garnett to Thomas Wolfe, and Robert Lowell to Dylan Thomas. The student cry to which the Apollonian spirit is responsive is "Teacher, tell it like it is," even when the cry comes from Dionysian lips.

Surely Adolf Hitler was one of the most able Dionysian teachers of recent times, and the blackness of his art was sinister. But all the performing arts are "black arts" because they depend on establishing an empathy, a resonance. The role of their practitioners is first to bemuse, then to engage, and finally to enlist their audiences.

"The Wedding Guest sat on a stone:/ He cannot choose but hear." That is magic; and not all magic is sinister, even though all of it is black. The

best magicians have always obtained their results
by means of which they are more or less aware but
which they cannot teach to others. There are
courses in acting, conducting, teaching—but
Merlin offers none in witchcraft, and the Sorcerer's
Apprentice came a cropper.

To place teaching among the black arts is
merely to classify it. It, like its fellows, is an art
in its own right. Whether or not it employs lectures, it is not the art of acting. But making its
membership in the arts explicit focuses attention
on some worthy habits that all the arts share.

In particular, the arts are aware of one another;
they look at one another out of the corners of
their eyes. One pleasant, if somewhat trivial,
result of this mutual inspection is their habit of
making graceful references to one another. There
are many paintings of musicians performing
music, and the tables are turned in Moussorgski's
"Pictures at an Exhibition." "L'Apres-midi d'un
faune" we know better as Debussy's music than
as Mallarmé's earlier poem.[5]

Much more important is the sharing of discovery. Just as the sciences, the arts make discoveries, and a discovery in one art can sometimes
be transmuted to serve a purpose in another.
Surrealism, popularly regarded as a painting
movement, was invented by the poet André
Breton, who directed its literary manifestations.
In the English pre-Raphaelite movement of the
last century, and in the French Dada movement
of this century, the interaction between poets
and painters was so close that it would be hard
to say which were the prime discoverers.

It may be that teaching—even physics teaching—has already made artistic discoveries that
are usable in the other arts. I have been told of
an eminent European physicist who made a practice of teaching the introductory physics course
in his university. As he lectured, scores of little
things were successively happening at the hands
of gremlins directed from a console on his lectern—
a fusillade of tiny events relating to his remarks.

Only now have visual artists rediscovered that
such goings on make an art form. Only now is

---

[5] And better than as Varèse's later parody, "L'Apres-
midi d'un Ford," octet for piccolos with solo Klaxon.

McLuhan erecting the assault upon all senses at once into an artistic communications theory. For teaching has not yet been noticed as an art by other artists, and their sidelong glances go elsewhere. And the glances of teachers, in their professional capacity, seldom stray to the other arts. Only painfully, belatedly, and in loneliness do they rediscover some of the artists' oldest discoveries.

To dispel this isolation, perhaps a small entering wedge would be participation in graceful reference. For example, I suspect that Hieronimus Bosch's paintings of Hell are full of elementary physics—levers, pulleys, and what not. His is a Hell, as an artist has remarked, to which nobody can imagine himself going: The analysis would be undistracted by strong emotion.

Consider, for another example, the opening musical phrase of the familiar French folksong, "Il était une bergère." When you take the words and the beat away, the bare succession of notes has time-reversal symmetry. Rhythm and meaning contribute "time's arrow": What is happening to the entropy of the situation then?

But interartistic reference degenerates into gimmickry if it is not graceful. Finding resemblances between "modern art" and the pictures taken by an electron microscope is amusing, but so is the old pastime of finding rabbits in cloud formations. He who adheres to "tell it like it is" has an allergy to the spurious. He who signs up for a course in "physics for poets" wants a course in physics, not poetry.

Again, much more important than these peripheral references is the sharing of discovery. Examine two large and ancient discoveries in the arts, with abiding application to all of them, including teaching. The first is the uses of *scale*, a discovery of the visual arts. The second is the uses of *repetition*, a discovery of the auditory arts.[6]

Turning to the visual arts and their lesson of

---

[6] Here for simplicity, I class literature as an auditory art. It is much more than that, of course. But Reuben Brower, deploring the ascendency of the "rapid reading" movement, reminds us that good writing, even if it is only expository, "sounds." Gibbon paced the floor as he wrote *The Decline and Fall of the Roman Empire*, reciting its sentences to be sure that they sounded.

scale, begin with the Roman portrait busts, carved through several centuries after Rome became an empire. Here was a remarkably perfected art form. Mounted at eye level in a museum, those people meet us as we enter the room . . . and they are the people that we see every day. We know them well—thoughtful, worried, responsible, arrogant. Added to superb craftsmanship, there is one controlling reason for the immediacy of these heads: They are almost life-size.

Heroic sculpture, no matter in what style, can never have such a uncanny immediacy. And it isn't intended to. These people, on horseback or wherever, are heroes. In legend they are larger than life; so must they be in effigy. The people carved on the side of Mt. Rushmore under the direction of Gutzon Borglum must be very important indeed.

Conversely the small plaster busts of famous men trivialize them, and carry neither dignity nor conviction. In point of fact, none of us little chaps could live day by day in equality with the familiar bust of Beethoven: It would shame us. The row of busts on the mantlepiece serves (say it softly) only to advertise to ourselves and our guests that we are cultured folk. Reduced to the size of jewelry, effigies acquire new interest, but purely one in craftsmanship.

There is a widely distributed little piece of moralistic sculpture showing three monkeys with their paws over their ears, their eyes, and their mouth. Hear no evil, see no evil, speak no evil. But the piece is always too small to carry conviction, and so its owner continues to hear, see, and speak evil anyway.

In short, the size of man is unity on the number-line of the visual arts. "Man is the measure of all things," said Plato. Notice the difference in the emotional impact, so to speak, of the two polyhedra in Fig. 1, one larger and one smaller than the head of a man.

You may protest, "These polyhedra are geometrically similar, and therefore identical for my purpose." But you are teaching Coleridge, whose sensitivities are sharper than yours in some areas, albeit less sharp in others. If you would teach him your way of abstraction—and probably you would, for it is at the heart of physics—do it slowly and explicitly.

I have heard of a game invented to teach very young children the beginnings of the idea of a number base. It employed white, red, and gray counters: Three white counters were worth one red counter, and three red were worth one gray. The game was a failure. On looking into why, it

Fig. 1. Scaled up by a factor of 4, Louis Poinsot's *great icosahedron* becomes 16 times as important to the eye, and 64 times as important to the kinesthetic imagination.

emerged that all right-thinking people know a red thing is worth more than a gray thing. After the color assignments were interchanged, the game went well.

I made a polyhedron consisting of three interpenetrating cubes, and I colored the exposed parts of the cubes red, white, and blue. To you and me the colors are alpha, beta, and gamma. I'm afraid some of my nonmathematical friends think I am suggesting that the inhabitants of the United States of America are all squares.

A little pendulum is very cute, ticking off the seconds. A big one, a really long one, is majestic and at first rather mysterious. After traveling in one direction for awhile, it stops and turns around, even though you don't do anything to it. Perhaps Newton's first law is "important"? Then level the longest lecture table you have, cut a big brick of dry ice, and start it moving slowly. "Hey, the guy's right! It *does* go on forever."

As for *repetition*, any teacher already knows one rudimentary and relevant fact about it. He seldom learned anything himself the first time he heard it and, remembering that, he doesn't expect his students to do so either. But, at the same time, repetition is a bore, to him and to them. So how does he go about the job of repeating the things to be learned? And does he do it in the same way for the big things and the little things? For light and guidance and style, I suggest that he examine some musical and poetical structures. The way to examine them is to listen.[7]

Music has *passacaglias*, in which a tune is slowly repeated, over and over, to form a ground bass for a harmonic and contrapuntal superstructure. It has *canons*—simple nursery rounds such as "Frère Jacques," more elaborate canons such as Bach's—in which one voice presents a tune and others take it up in overlapping repetition. It has *fugues*, where a tune, after successive presentation by several voices in two related keys, is treated more freely and may even appear upside down or twice as fast. It has "Variations on a Theme by So-and-So" in which, after a simple statement, a tune is elaborated and ornamented in ways that delight by their increasing ingenuity.

Poetry has rhyme and meter—simple kinds of repetition—and assonance, which is subtler. It has repetitious structures such a ballade, rondeau, and triolet. It has refrain and envoy. It may induce surprise or shock by the sudden reappearance of an utterance. Examine Conrad Aiken's "Priapus and the Pool" for the many ways in which that poem presses antiphony and repetition into service.

---

[7] To the interviewer who asked her what she thought of modern art, Gertrude Stein replied, "I like to look at it."

As you listen to poetry, sharpen your ear. For a silly example, take the limericks of Edward Lear. Aficionados of the limerick often find regrettable the bathos in his habit of repeating his first line for his last. But note the little twist in: There was an old person of Wick/ Who said "tick-a-tick, tick-a-tick,/ Chick-a-bee, chick-a-baw,"/ And he said nothing more,/ That laconic old person of Wick.[8]

I have heard of a course of lectures, delivered to graduate students at Harvard by a German refugee professor, in which for several weeks he started each lecture in the same way. "Last time we were talking about the Michelson–Morley experiment. Which is this!" And he turned to the blackboard and put up the same diagram. But then he sank his teeth in the experiment from one angle, and next time he grabbed and shook it from another. It was a style that epitomized the way physics itself is actually done. One, at least, of his auditors has never forgotten it.

There are larger auditory art forms. Music has symphonies, whose themes are repeated and varied for larger purposes. The last movement of Beethoven's Third Symphony takes a free theme-and-variations form, in which the musical possibilities in a simple succession of four notes are developed through humor, to triumph, to sadness. In a cyclical form such as that of César Franck's only symphony, the themes of earlier movements all reappear in the last. And there is Richard Wagner's opera, "Die Meistersinger," which is about a fine tune: The tune appears in fragments and in travesties, and only near the end does it receive full, final, authentic statement.

And correspondingly there are larger matters of physical concern. Newton's third law is pretty big stuff. It is a theme whose implications and uses do not emerge on a first statement. Its development for a student surely demands a symphonic or operatic scope: Perhaps Beethoven, Franck, or Wagner can help to suggest style.

And so, teacher, if you would teach physics to Coleridge or to anybody else, remember that you are an artist, and behave as artists do. Paint your

---

[8] Remember that "laconic" doesn't mean "taciturn." It means saying a great deal in very few words. In this example, so does Lear.

picture, choosing from the rich palette laid out for you by the physicists those colors that suit at once your client's intellectual décor and your own high purpose. Wear the label of artist proudly, and be not overwhelmed by its enormous implications. The great artists of the past were mostly talented craftsmen serving large causes. The painters of the Italian Renaissance were hired to expound the Roman Catholic version of the story of Jesus and the Roman Catholic vision of the divine sanctions. Bach composed and conducted his cantatas, the magnificent central corpus of his work, in the service of Lutheranism. You too are in the fortunate position that they were in, burnishing your tools and lavishing your craftsmanship in the service of something larger than yourself, in which you believe.

Chapter One

PEOPLE IN PHYSICS

What is your prototype of the scientist? Is it the mad professor, benevolent magician, or uninspired pedant as portrayed by Arthur Koestler in "Three Character-Types"? You may apply his descriptions to Galileo, Newton, Franklin, Faraday, and Maxwell, about whom Koestler has written brief sketches, as well as to your physics instructor or other scientists of your acquaintance. The published scientific work of these men does not reveal all facets of their personalities.

Much greater detail of Newton's extrascientific work is given in the tercentenary lecture "Newton, The Man" prepared by Lord Keynes. Keynes made a hobby of collecting Newton's "magical" papers, and he presents a fascinating account of the abstruse questions that possessed the mind of the genius. During the quarter century of intense activity from 1663 to 1688, Newton appears to fit most closely Koestler's "mad professor" type. The description of his scientific contributions in Chapters 14 and 15 of your text is difficult to reconcile with these revelations.

Albert Einstein, whose genius rivaled Newton's and whose work superseded Newton's in many important respects, is rightly seen as "benevolent magician." The selection about Einstein's early years (leading up to his statement of the Special Theory of Relativity in 1905) is taken from the first of three programs prepared by the British Broadcasting Corporation and originally suggested by a group of Israeli scientists. The programs were produced under the general editorship of Dr. Gerald Whitrow and included contributions from Einstein's son and many of Einstein's associates.

Niels Bohr, a leader with Einstein in the creation of modern quantum and relativity physics (see Chapter 8 of Introductory Physics, A Model Approach), was also a "benevolent magician," but his personality emerges only partly from the one chapter of Ruth Moore's biography included here. What you can see, however, is a picture of the intense critical review to which the concept of complementarity was subjected by its originators. Here Einstein played the role of sceptic rather than of innovator.

In the last selection of this chapter, you will get an additional view of the famous physicists described in earlier articles. Try to identify the individuals from the clues given, before you read their names! Raymond J. Seeger is particularly concerned with their childhood and youth, however, and not with their scientific work. You may like to speculate about the influence these men's early lives and varied fortunes had on their later creative accomplishments.

About the Contributors

ARTHUR KOESTLER (1905-   ) is an author and playwright who resides in London. He has traveled widely, fought in the Spanish Civil War, and has expressed a strongly anti-Communist philosophy in many writings. Mr. Koestler's early training as a scientist in Vienna, combined with his broad human experience, are the background from which he has produced The Act of Creation.

JOHN MAYNARD KEYNES (1883-1946), Lord Keynes of Tilton, was one of the most famous economists of the first half of the twentieth century. He revolutionized economic thinking by stressing the importance of a broad distribution of wealth, even at the expense of governmental deficits in times of depression. Lord Keynes taught at Cambridge for more than thirty years, and collected information about his illustrious Cambridge predecessor Isaac Newton as a hobby.

GERALD J. WHITROW (1912-   ) is Reader in Applied Mathematics at the Imperial College, University of London. He is also President of the British Society for the History of Science. He has written papers and books on relativity, cosmology, and the subject of time, and has given many broadcast talks on these subjects as well.

ANDRE D. H. MERCIER (1913-   ) is Head of the Department of Theoretical Physics at Berne University in Switzerland. He is a geophysicist and Secretary-General of the International Committee on General Relativity and Gravitation.

BERNARD MAYES (1929-   ) is a veteran broadcaster and the resident California Correspondent for the British Broadcasting Corporation. His commentaries and interviews on American affairs are broadcast around the world. He was educated at Cambridge University and is now an American citizen, serving on the Board of Directors of National Public Radio in the United States.

HANS EINSTEIN (1904-   ) is Professor of Hydraulic Engineering at the University of California at Berkeley. His special field is tidal flow and the behavior of sediments, an important area which has led to service with several government agencies. Professor Einstein is the elder son of Albert Einstein.

BERTRAND RUSSELL (1872-1970), Earl Russell, was a British mathematician and philosopher. His mathematical studies were directed toward the establishment of complete logical systems. Earl Russell was a militant pacifist, whose outspokenness involved him in numerous controversies. In 1950 he won the Nobel Prize for literature.

SIR KARL R. POPPER (1902-   ) was Professor of Logic and Scientific Method at the University of London until his recent retirement. He has written many books on the philosophy of science, has lectured widely, and has earned many distinctions for his work.

RUTH MOORE is an author who has written many books about science and scientists for the general public. Among them are Charles Darwin; A Great Life in Brief (1955) and The Coil of Life: The Story of the Great Discoveries in the Life Sciences (1961). Miss Moore has worked as reporter, correspondent, and feature writer for several newspapers. Her deep concern about housing, public welfare, and urban renewal has made her a specialist on these questions

RAYMOND J. SEEGER (1906-   ) is Senior Staff Associate for Research at the National Science Foundation. Dr. Seeger has taught college physics and is especially interested in the relation of culture and physics. During World War II he studied problems related to shock waves and explosives. For several years he directed research programs in these areas at government and university laboratories.

Three Character-Types

Let me revert for a moment to our starting point, the triptych of creative activities.

In folklore and popular literature the Artist is traditionally represented as an inspired dreamer—a solitary figure, eccentric, impractical, unselfish, and quixotic.

His opposite number is the earthy and cynical Jester—Falstaff or Sancho Panza; he spurns the dreamer, refuses to be taken in by any romantic nonsense, is wide-awake, quick to see his advantage and to get the better of his fellows. His weapons range from the bludgeon of peasant cunning to the rapier of irony; he always exercises his wits at the expense of others; he is aggressive and self-asserting.

In between these antagonistic types once stood the Sage who combined the qualities of both: a sagacious dreamer, with his head in the clouds and his feet on the solid earth. But his modern incarnation, the Scientist, is no longer represented by a single figure in the waxworks of popular imagination; instead of one prototype, we had better compose three.

The first is the Benevolent Magician, whose ancestry derives from the rain-making Shamans and the calendar-making Priest-Astronomers of Babylon. At the dawn of Greek science we find him assuming the semi-mythical figure of Pythagoras, the only mortal who could hear with ears of flesh the music made by the orbiting stars; and from there onward, every century created its own savant-shamans whom it could venerate—even throughout the Dark Ages of science. The first millennium was seen in by Sylvester II, the 'Magician Pope', who reinstated the belief that the earth was round. The Jews had their Maimonides, the Arabs their Alkhazen, Christendom the Venerable Bede, before St. Thomas Aquinas and Albert the Great revived the study of nature.

255

From the Renaissance on there is an uninterrupted procession of magicians whose names were legends, admired and worshipped by a public which had only the vaguest notion of their achievements: Paracelsus, Tycho on his Sorcerer's Island, Galileo with his telescope, Newton who brought the Light, Franklin who tamed the thunderbolt, Mesmer who cured by magnetism; Edison, Pasteur, Einstein, Freud. The popular image of the Magician has certain features in common with that of the Artist: both are unselfishly devoted to lofty tasks—which frequently overlapped in the *uomo universale* of the Renaissance.

The second archetype is the 'Mad Professor' who, in contrast to the former, practises black instead of white magic for the sake of his own aggrandizement and power. Eudoxus jumped into the crater of Etna to gain immortality; Paracelsus's rival, Agrippa, was allied to the devil in the shape of an enormous black poodle; the Anatomists were allied to body-snatchers for their sinister purposes. The alchemists distilled witches' brews; electric rays became a favourite delusion in persecution manias; vivisection, and even compulsory vaccination, became symbols of the scientist's blasphemous presumption and cruelty. The Mad Professor—either a sadist or obsessed with power—looms large in popular fiction from Jules Verne's Captain Nemo and H. G. Wells' Dr. Moreau to Caligari, Frankenstein, and the monsters of the horror-comics. He is a Mephistophelian character, endowed with caustic wit; he spouts sarcasm, a sinister jester plotting to commit some monstrous practical joke on humanity. His place in the waxworks is next to the malicious satirist's, as the Benevolent Magician's is next to the imaginative Artist's.

The last of the three figures into which the popular image of the scientist has split occupies the centre space and is of relatively recent origin: the dry, dull, diligent, pedantic, uninspired, scholarly bookworm or laboratory worker. He is aloof and detached, not because he has outgrown passion but because he is devoid of temperament, desiccated, and hard of hearing—yet peevish and petulant and jealous of anybody who dares to interfere with his crabbed little world. This imaginary type probably originates with the Schoolmen of the period of decline, whom Erasmus lampooned: 'They smother me beneath six hundred dogmas; they are surrounded with a bodyguard of definitions, conclusions, corrolaries, propositions explicit and propositions implicit; they are looking in utter darkness for that which has no existence whatsoever.'

Swift satirized the type in *Gulliver in Laputa;* then Goethe in his *Famulus* Wagner: *Mit Eifer hab' ich mich der Studien beflissen—Zwar weiss ich viel, doch möcht' ich alles wissen.* 'Thanks to my diligence, my wisdom is growing—If I but persevere I shall be all-knowing.' His modern incarnations are the Herr Professor of German comedy, and the mummified dons of Anglo-Saxon fiction. At his worst, he incarnates the pathological aspects in the development of science: rigidity, orthodoxy, snowblindness, divorce from reality. But the patience and dogged endurance of the infantrymen of science are as indispensable as the geniuses who form its spearhead. 'The progress of science', Schiller wrote, 'takes place through a few master-architects, or in any case through a number of guiding brains which constantly set all the industrious labourers at work for decades.'[1] That the industrious labourers tend to form trade unions with a closed-shop policy and restrictive practices, is an apparently unavoidable development. It is no less conspicuous in the history of the arts: the uninspired versifiers, the craftsmen of the novel and the stage, the mediocrities of academic painting and sculpture, they all hang on for dear life to the prevailing school and style which some genius initiated, and defend it with stubbornness and venom against heretic innovators.

Thus we now have five figures facing us at our allegorical Madame Tussaud's. They are from left to right: the malicious Jester; the Mad Professor with his delusions of grandeur; the uninspired Pedant; the Benevolent Magician; and the Artist.

At the moment only the three figures in the centre concern us. If we strip them of the gaudy adornments which folklore and fiction bestowed upon them, the figure of the Black Magician will turn out to be an archetypal symbol of the *self-assertive element* in the scientist's aspirations. In mythology, this element is represented by the Promethean quest for omnipotence and immortality; in science-fiction it is caricatured as a monstrous lusting for power; in actual life, it appears as the unavoidable component of competitiveness, jealousy, and self-righteousness in the scientist's complex motivational drive. 'Without ambition and without vanity', wrote the biologist Charles Nicolle, 'no one would enter a profession so contrary to our natural appetites.'[2] Freud was even more outspoken: 'I am not really a man of science, not an observer, not an experimenter, and not a thinker. I am by temperament nothing but a conquistador ... with the curiosity, the boldness, and the tenacity that belong to that type of person.'[3]

The unassuming figure of the Pedant in the centre of the waxworks

7

is an indispensable stabilizing element; he acts as a restraining influence on the self-asserting, vainglorious conquistadorial urges, but also as a sceptical critic of the inspired dreamer on his other side.

This last figure, the White Magician, symbolizes the *self-transcending element* in the scientist's motivational drive and emotional make-up; his humble immersion into the mysteries of nature, his quest for the harmony of the spheres, the origins of life, the equations of a unified field theory. The conquistadorial urge is derived from a sense of power, the participatory urge from a sense of oceanic wonder. 'Men were first led to the study of natural philosophy', wrote Aristotle, 'as indeed they are today, by wonder.'[4] Maxwell's earliest memory was 'lying on the grass, looking at the sun, and *wondering*'. Einstein struck the same chord when he wrote that whoever is devoid of the capacity to wonder, 'whoever remains unmoved, whoever cannot contemplate or know the deep shudder of the soul in enchantment, might just as well be dead for he has already closed his eyes upon life'.[5]

This oceanic feeling of wonder is the common source of religious mysticism, of pure science and art for art's sake; it is their common denominator and emotional bond.

'You cannot help it, Signor Sarsi, that it was granted to me alone to discover all the new phenomena in the sky and nothing to anybody else.' The most conspicuous feature in the character of Galileo (1564–1642) and the cause of his tragic downfall was vanity—not the boisterous and naïve vanity of Tycho, but a hypersensitivity to criticism combined with sarcastic contempt for others: a fatal blend of genius plus arrogance minus humility. There seems to be not a trace here of mysticism, of 'oceanic feeling'; in contrast to Copernicus, Tycho, and Kepler, even to Newton and Descartes who came after him, Galileo is wholly and frighteningly modern in his consistently mechanistic philosophy. Hence his contemptuous dismissal in a single sentence of Kepler's explanation of the tides by the moon's attraction: 'He [Kepler] has lent his ear and his assent to the moon's dominion over the waters, to occult properties and such like *fanciullezze*.' The occult little fancy he is deriding is Kepler's anticipation of Newtonian gravity.

Where, then, in Galileo's personality is the sublime balance between self-asserting and self-transcending motives which I suggested as the true scientist's hallmark? I believe it to be easily demonstrable in his writings on those subjects on which his true greatness rests: the first discoveries with the telescope, the foundations of mechanics, and of a truly experimental science. Where that balance is absent—during the tragic years 1613–33, filled with poisonous polemics, spurious priority claims, and impassioned propaganda for a misleadingly oversimplified Copernican system—in that sad middle period of his life Galileo made no significant contribution either to astronomy or to mechanics. One might even say that he temporarily ceased to be a scientist—precisely because he was entirely dominated by self-asserting motives. The opposite kind of imbalance is noticeable in Kepler's periods of depression, when he entirely lost himself in mystic speculation, astrology, and number-lore. In both these diametrically opposed characters, unsublimated residues of opposite kind temporarily dominated the field, upsetting the equilibrium and leading to scientific sterility.

But in the balanced periods of Galileo, the eighteen happy years in Padua in which most of his epoch-making discoveries in the study of motion were made, and in the last years of resignation, when he completed and revised the *Dialogue Concerning Two New Sciences*—in these creative periods we seem to be dealing with a different kind of person, patiently and painstakingly experimenting and theorizing on the motions of the pendulum; on the free fall and descent along an inclined plane of heavy bodies; on the flight of projectiles; the elasticity,

9

cohesion, and resistance of solid bodies, and the effects of percussion on them; on the buoyancy of 'things which float on the water', and a hundred related matters. Here we have a man absorbed in subjects much less spectacular and conducive to fame than the wonders of the Milky Way and the arguments about the earth's motion—yet delighting in his discoveries, of which only a select few friends and correspondents were informed; delighting in discovery for discovery's sake, in unravelling the laws of order hidden in the puzzling diversity of phenomena.

That order was for Galileo, as it was for Kepler, a mathematical order: 'The book of nature is written in the mathematical language. Without its help it is impossible to comprehend a single word of it.' But unlike Kepler and the Pythagoreans, Galileo did not look at the 'dance of numbers' through the eyes of a mystic. He was interested neither in number-lore nor in mathematics for its own sake—almost alone among the great scientists of his period, he made no mathematical discoveries. Quantitative measurements and formulations were for Galileo simply the most effective tools for laying bare the inherent *rationality of nature*. The belief in this rationality (and in the rationality of nature's creation, the human mind) was Galileo's religion and spiritual salvation—though he did not realize that it was a religion, based on an act of faith.

His revolutionary methods of proving the rationality of the laws governing the universe was later called 'experimental philosophy'— and even later, by the much narrower terms 'experimental science' or 'empirical science'. It was a fertile combination of experimenting and theorizing, which had been tentatively used by some of Galileo's precursors since the fourteenth century—but it was Galileo who elevated it to a modern technique and a philosophical programme. It was a monumental bisociation of the valid elements in Greek thought, transmitted by the Schoolmen (and particularly by the Occamists) on the one hand, and of the experimental knowledge of engineers, artisans, and instrument-makers on the other. The *Dialogue Concerning Two New Sciences* characteristically opens with a most unusual suggestion by Salviati (Galileo's mouthpiece): that, as a philosopher, he had much to learn from mechanics and craftsmen.

> *Salviati:* The constant activity which you Venetians display in your famous arsenal suggests to the studious mind a large field for investigation, especially that part of the work which involves mechanics;

for in this department all types of instruments and machines are constantly being constructed by many artisans, among whom there must be some who, partly by inherited experience and partly by their own observations, have become highly expert and clever in explanation.

Sagredo: You are quite right. Indeed, I myself, being curious by nature, frequently visit this place for the mere pleasure of observing the work of those who, on account of their superiority over other artisans, we call 'first-rank men'. Conference with them has often helped me in the investigation of certain effects including not only those which are striking, but also those which are recondite and almost incredible.

We are reminded of Pythagoras visiting the blacksmith's shop to discover the secret of vibrating chords—to learn from those dark, sweaty, and ignorant men about the harmony of the spheres. This is the point where *hubris* yields to humility; in his best and happiest moments, Galileo achieves not only this transition, but is also transformed from a scientist into a poet. In the midst of his formidable polemical onslaught on the Platonist dualism of despair—which contrasted the perfect, immutable, crystalline heavens to the earthy corruption of generation and decay—his imagination and language suddenly grow wings:

Sagredo: I cannot without great wonder, nay more, disbelief, hear it being attributed to natural bodies as a great honour and perfection that they are impassible, immutable, inalterable, etc.: as, conversely, I hear it esteemed a great imperfection to be alterable, generable, mutable, etc. It is my opinion that the Earth is very noble and admirable by reason of the many and different alterations, mutations, generations, etc., which incessantly occur in it. And if, without being subject to any alteration, it had been all one vast heap of sand, or a mass of jade, or . . . an immense globe of crystal, wherein nothing had ever grown, altered, or changed, I should have esteemed it a wretched lump of no benefit to the Universe, a mass of idleness. . . . What greater folly can be imagined than to call gems, silver, and gold noble and earth and soil base? . . . If there were as great a scarcity of earth as there is of jewels and precious metals, there would be no king who would not gladly give a heap of diamonds and rubies . . . to purchase only so much earth as would suffice to plant a jessamine in a little pot or to set a tangerine in it, that he might see it

11

sprout, grow up, and bring forth goodly leaves, fragrant flowers, and delicate fruit. . . . These men who so extol incorruptibility, inalterability, etc., speak thus, I believe, out of the great desire they have to live long and for fear of death, not considering that, if men had been immortal, they would not have had to come into the world. These people deserve to meet with a Medusa's head that would transform them into statues of diamond and jade that so they might become more perfect than they are.

In another work, Galileo wrote a charming and profound allegory on the motives, methods, and limitations of the 'experimental philosophy' which he had created. The work is *Il Saggiatore*, 'The Assayer' —which has only recently been translated into English, presumably because most of it consists of a querulous, scientifically worthless polemics against the Jesuit scholar Grassi on the subject of comets (which Galileo insisted on treating as optical illusions—largely because Tycho and Grassi held the opposite views). Yet hidden in this nasty bunch of nettles are flowers of rare beauty:

> Once upon a time, in a very lonely place, there lived a man endowed by nature with extraordinary curiosity and a very penetrating mind. For a pastime he raised birds, whose songs he much enjoyed; and he observed with great admiration the happy contrivance by which they could transform at will the very air they breathed into a variety of sweet songs.
> One night this man chanced to hear a delicate song close to his house, and being unable to connect it with anything but some small bird he set out to capture it. When he arrived at a road he found a shephered boy who was blowing into a kind of hollow stick while moving his fingers about on the wood, thus drawing from it a variety of notes similar to those of a bird, though by a quite different method. Puzzled, but impelled by his natural curiosity, he gave the boy a calf in exchange for this flute and returned to solitude. But realizing that if he had not chanced to meet the boy he would never have learned of the existence of a new method of forming musical notes and the sweetest songs, he decided to travel to distant places in the hope of meeting with some new adventure.

Subsequently, the man discovered that there are many other ways of producing musical notes—from strings and organs, to the swift vibrations on the wings of mosquitoes and the 'sweet and sonorous shrilling of crickets by snapping their wings together, though they cannot fly at all'. But there was an ultimate disappointment waiting for him:

> Well, after this man had come to believe that no more ways of forming tones could possibly exist ... when, I say, this man believed he had seen everything, he suddenly found himself once more plunged deeper into ignorance and bafflement than ever. For having captured in his hands a cicada, he failed to diminish its strident noise either by closing its mouth or stopping its wings, yet he could not see it move the scales that covered its body, or any other thing. At last he lifted up the armour of its chest and there he saw some thin hard ligaments beneath; thinking the sound might come from their vibration, he decided to break them in order to silence it. But nothing happened until his needle drove too deep, and transfixing the creature he took away its life with its voice, so that he was still unable to determine whether the song had originated in those ligaments. And by this experience his knowledge was reduced to diffidence, so that when asked how sounds were created he used to answer tolerantly that although he knew a few ways, he was sure that many more existed which were not only unknown but unimaginable.

Hubris is temporarily submerged by humility. Galileo was the first of a race of modern experimental scientists convinced of the infallibility of their 'exact empirical methods'; in fact he created the type. It comes as a surprise to hear him talk about things 'not only unknown but unimaginable'. But this ultimate modesty, derived from a sense of wonder close to mysticism, is found in all great scientists—even if hidden by an arrogant façade, and allowed to express itself only on rare occasions.

About Kepler I have said enough, in this book and elsewhere, to show that mysticism was the mainspring of his fantastically laborious life—starting with the analogy between God the Father and the Sun, continued in his lifelong conviction that the universe was built around the frames of the five Pythagorean solids, and that the planetary motions were regulated by the laws of musical harmony. But his mystic convictions, and the disarmingly child-like streak in his character, did not

13

prevent him from casting horoscopes for money—however much he despised himself for it; from indulging in naïve snobbery, and quarrelling like a fish-wife with the overbearing Tycho. His vanity had a perverse twist: he was very proud of himself when his astrological forecasts of a cold spell and an invasion by the Turks came true; but towards his real discoveries he was completely indifferent, and he was astonishingly devoid of professional jealousy. He naïvely expected the same of other astronomers; and when Tycho's heirs delayed publication of his priceless collection of observational data, Kepler simply stole the material to put it to proper use—his ethics did not include respect for private property in Urania's domains.

When Kepler had completed the foundations of modern astronomy by his Third Law, he uttered a long Eureka cry:

> The heavenly motions are nothing but a continuous song for several voices (perceived by the intellect, not by the ear); a music which, through discordant tensions, through sincopes and cadenzas, as it were (as men employed them in imitation of those natural discords), progresses towards certain pre-designed, quasi six-voiced clausuras, and thereby sets landmarks in the immeasurable flow of time. It is, therefore, no longer surprising that man, in imitation of his creator, has at last discovered the art of figured song, which was unknown to the ancients. Man wanted to reproduce the continuity of cosmic time within a short hour, by an artful symphony for several voices, to obtain a sample test of the delight of the Divine Creator in His works, and to partake of his joy by making music in the imitation of God.

Here we have the perfect union of the two drives: the vain-glorious ego purged by cosmic awareness—*ekstasis* followed by *katharsis*.

Newton, Monster and Saint

From the end of the seventeenth century onward the scene becomes too crowded for a systematic inquiry into individual motivations; however, I have said enough to suggest the basic pattern—and though the character of the times changed, that pattern remained essentially the same.

Look at Newton, for instance: he has been idolized and his character bowdlerized to such an extent (above all in the Victorian standard biography by Brewster) that the phenomenal mixture of monster and saint out of which it was compounded was all but lost from sight. On the one hand he was deeply religious and believed—with Kepler and Bishop Usher—that the world had been created in 404 B.C.; that the convenient design of the solar system—for instance, all planetary orbits lying in a single plane—was proof of the existence of God, who not only created the universe but also kept it in order by correcting from time to time the irregularities which crept into the heavenly motions —and by preventing the universe from collapsing altogether under the pressure of gravity. On the other hand, he fulminated at any criticism of his work, whether justified or not, displayed symptoms of persecution mania, and in his priority fight with Leibniz over the invention of the calculus he used the perfidious means of carefully drafting in his own hand the findings, in his own favour, of the 'impartial' committee set up by the Royal Society. To quote M. Hoskin:

> No one supposes that the committee set up by the Royal Society of which Newton had then been president for several years, was impartial. But we can only realize the extent of Newton's share in its conclusions when we examine a much-corrected draft summary of what were to be the findings of the committee. The draft is written in Newton's own hand, and it is fascinating to watch Newton debating with himself whether the committee ought to say 'We are satisfied that he [Newton] had invented the method of fluxions before' 1669, or whether it would sound better if they said 'We find that he invented the method of fluxions before' 1669; or deciding that to say 'We are satisfied that Mr. Newton was the first author of this method' was too terse, and that several more lines of explanation ought to be inserted before the conclusion 'for which reason we reckon Mr. Newton the first inventor'.

Here is pettiness on a heroic scale—combined with a heroic vision of the universe worked out in minute detail: in other words, the mixture as before.

The Mysticism of Franklin

As we move on into the eighteenth century the towering genius of Benjamin Franklin sticks out of it like his lightning rod. Printer,

15

journalist, pamphleteer, politician, wire-puller, diplomat, and states-
man; pioneer of electricity, founder of the physics of liquid surfaces,
discoverer of the properties of marsh gas, designer of the *chevaux de
frise* which halted the advance of the British fleet on the Delaware,
inventor of bifocal spectacles and of improved fireplaces, advocate of
watertight bulkheads on ships and of chimney-shafts for the ventila-
tion in mines—the list could be continued. And yet this 'first civilized
American', as one of his biographers called him,　for all his incompar-
able clarity of thought and lucidity of style, had formed his metaphysical outlook at the age of sixteen when he read a book by Tryon,
a member of the group of British Pythagoreans. The members of this
sect were chiefly known for their vegetarianism because, like the
ancient Brotherhood, they believed in the transmigration of souls and
wished to avoid the risk of feasting on some reincarnation of a human
being. Franklin became a convert to vegetarianism and believed in
transmigration to the end of his life. At the age of twenty-two he
composed a Pythagorean epitaph for himself; at the age of eighty-four,
the year of his death, he ordered that it should appear, unchanged, on
his tomb. It reads:

The Body
Of
BENJAMIN FRANKLIN
Printer
(Like the Cover of an Old Book
Its Contents Torn Out
And Stript of its Lettering and Gilding)
Lies Here, Food for Worms.
But the Work Shall Not Be Lost
For It Will (As He Believed) Appear Once More
In a New and More Elegant Edition
Revised and Corrected
By
The Author

His conviction that souls are immortal, that they cannot be destroyed and are merely transformed in their migrations led him, by
way of analogy, to one of the first clear formulations of the law of the
conservation of matter. The following quotations will make the
connection clear:

16

The power of man relative to matter seems limited to the dividing it, or mixing the various kinds of it, or changing its form and appearance by differing compositions of it, but does not extend to the making or creating of new matter, or annihilating the old.

This was written when he was seventy-eight. The following was written one year later:

I say that when I see nothing annihilated, and not even a drop of water wasted, I cannot suspect the annihilation of souls, or believe that He will suffer the daily waste of millions of minds ready made that now exist, and put Himself to the continual trouble of making new ones. Thus finding myself to exist in the world, I believe I shall, in some shape or other, always exist.

The argument seems to indicate that what one might call the principle of the 'conservation of souls' was derived from that of the 'conservation of matter'. But in fact it was the other way round. As Kepler had transformed the Holy Trinity into the trinity of Sun—Force—Planets, so in Franklin's case, too, a mystical conviction gave birth, by analogy, to a scientific theory. And could there be a more charming combination of man's vanity with his transcendental aspirations than to pray for a 'more elegant, revised, and corrected edition' of one's proud and humble self?

The Fundamentalism of Faraday

The nineteenth-century landscape is crowded with giants; I shall briefly comment on four of them. In the physical sciences Faraday and Maxwell are probably the greatest: Einstein, who ought to know, has put them on a par with Galileo and Newton; and Crowther, who wrote short biographies of both, makes the fine distinction of calling Faraday 'the greatest physicist of the nineteenth century' and Maxwell 'the greatest theoretical physicist of the nineteenth century'. To these let me add, from the biological sciences, Darwin and Pasteur, to make up a foursome.

Faraday, whom Tyndall described as 'the great mad child', was the most inhuman character of the four: the son of a sectarian blacksmith, self-taught, with a passionate temperament which was denied all

17

human outlets except religion and science. This was probably the cause of the protracted episode of mental disorder, comparable to Newton's, which began when he was forty-nine. Characteristic of the coyness of science historians is the *Encyclopaedia Britannica*'s reference to Faraday's clinical insanity: 'In 1841 he found that he required rest, and it was not till 1845 that he entered on his second great period of research.'

At thirty, shortly after his marriage—which remained childless—Faraday joined an extreme fundamentalist, ascetic sect, the 'Sandemanians', to which his father and his young wife belonged, and whose services he had attended since infancy. The Sandemanians considered practically every human activity as a sin—including even the Victorian virtue of saving money; they washed each other's feet, intermarried, and refused to proselytize; on one occasion they suspended Faraday's membership because he had to dine, by royal command, with the Queen at Windsor, and thus had to miss the congregation's Sunday service. It took many years before he was forgiven and re-elected an Elder of the sect.

In his later years Faraday withdrew almost completely from social contacts, refusing even the presidency of the Royal Academy because of its too worldly disposition. The inhuman self-denials imposed by his creed made Faraday canalize his ferocious vitality into the pursuit of science, which he regarded as the only other permissible form of divine worship.

The Metaphysics of Maxwell

James Clerk Maxwell was of an altogether different, balanced, and happy disposition. In his case, too, religious belief became a spur to scientific activity, but in more subtle ways. He was a double-faced giant: he completed the classical edifice of the Newtonian universe, but he also inaugurated the era of what one might call the 'surrealistic' physics of the twentieth century.

As Kepler had embraced the Copernican system 'for physical or if you prefer, metaphysical reasons', so Maxwell confessed that the theories of his later period were formed 'in that hidden and dimmer region where Thought weds Fact. Does not the way to it pass through the very den of the metaphysician, strewed with the remains of former explorers and abhorred by every man of science?'

The metaphysician in Maxwell had by that time long outgrown the crude materialism of mid-nineteenth-century science, and its equally crude forms of Christianity. Maxwell's religious beliefs were conceived in symbolic, almost abstract, terms; they compared to Faraday's fundamentalist creed as his abstract equations of the electro-magnetic field compare with the lines of force which to Faraday were 'as real as matter'. The connection between Maxwell's religious and scientific views is indeed just as intimate as in the case of Franklin or Kepler. I have mentioned before how, once he had arrived at his twenty general equations, Maxwell kicked away the scaffolding from under him—the physical model of vortices in the ether—and thus inaugurated the post-Newtonian era in physics, with its renunciation of all models and representations in terms of sensory experience.

There is a characteristic passage in one of his letters to his wife:

'I can always have you with me in my mind—why should we not have our Lord always before us in our minds. . . . If we had seen Him in the flesh we should not have known Him any better, perhaps not so well.' In another letter to his wife, he says that he had been re-reading Ephesians vi. This is not a very inspiring chapter, dealing with relations between parents and children, masters and servants; yet Maxwell comments: 'Here is more about family relations. There are things which have meanings so deep that if we follow on to know them we shall be led into great mysteries of divinity. If we reverence them, we shall even see beyond their first aspect a spiritual meaning. For God speaks to us more plainly in these bonds of our life than in anything that we can understand.'

J. G. Crowther—who, as an adherent of the Marxist philosophy of history can hardly be accused of mystic inclinations—remarks on this curious passage: 'Here Maxwell accepts material relationships with the belief that acquaintance with them will lead to spiritual understanding. He proceeds from the contemplation of material relationships to spiritual truth, from the model of the electro-magnetic field to the equations. The influence of the New Testament is seen also in hi sinterpretation of self-sacrifice. During the last years of his life, his wife was an invalid. He nursed her personally with the most assiduous care. At one period he did not sleep in a bed for three weeks, though he delivered his lectures and superintended the laboratory as usual. The modernity of Maxwell's science, and the antiquity of his sociology and religion appear incongruous. But it may be noted that though

19

his views on sociology and religion were antique, they were superior to those of nearly all his scientific contemporaries. He at least thought about these problems, and if he was unable to find modern answers to them, he learned enough of them to avoid the intellectual philistinism of his time.'

It was the time when Berthelot proclaimed: 'The world today has no longer any mystery for us'; when Haeckel had solved all his *Welträtsel* and A. R. Wallace, in his book on *The Wonderful Century*, declared that the nineteenth century had produced 'twenty-four fundamental advances, as against only fifteen for all the rest of recorded history'. The Philistines everywhere were 'dizzy with success'—to quote once more Stalin's famous phrase of 1932, when factories and power dams were going up at great speed while some seven million peasants were dying of starvation. It had indeed been a wonderful century for natural philosophy, but at its end moral philosophy had reached one of its lowest ebbs in history—and Maxwell was well aware of this. He was aware of the limitations of a rigidly deterministic outlook; it was he who, in his revolutionary treatment of the dynamics of gases, replaced mechanical causation by a statistical approach based on the theory of probability—a decisive step towards quantum physics and the principle of indeterminism. Moreover, he was fully aware of the far-reaching implications of this approach, not only for physics but also for philosophy: 'It is probable that important results will be obtained by the application of this the statistical method, which is as yet little known and is not familiar to our minds. If the actual history of Science had been different, and if the scientific doctrines most familiar to us had been those which must be expressed in this way, it is possible that we might have considered the existence of a certain kind of contingency a self-evident truth, and treated the doctrine of philosophical necessity as a mere sophism.'

Already at the age of twenty-four he had realized the limitations of materialist philosophy: 'The only laws of matter are those which our minds must fabricate, and the only laws of mind are fabricated for it by matter.' Twenty years later, at the height of his fame, he gave full rein to his hobby, satirical verse, to ridicule the shallow materialism of the Philistines. The occasion was the famous presidential address by John Tyndall to the British Association meeting in Belfast. Tyndall, a generous soul but a narrow-minded philosopher, attacked the 'theologians' and extolled the virtues of the brave new materialist creed. Maxwell's satire is still valid today:

20

In the very beginning of science,
    the parsons, who managed things then,
Being handy with hammer and chisel,
    made gods in the likeness of men;
Till Commerce arose, and at length
    some men of exceptional power
Supplanted both demons and gods by
    the atoms, which last to this hour.

From nothing comes nothing, they told us,
    nought happens by chance but by fate;
There is nothing but atoms and void,
    all else is mere whims out of date!
Then why should a man curry favour
    with beings who cannot exist,
To compass some petty promotion
    in nebulous kingdoms of mist? . . .

First, then, let us honour the atom,
    so lively, so wise, and so small;
The atomists next let us praise, Epicurus,
    Lucretius, and all;
Let us damn with faint praise Bishop Butler,
    in whom many atoms combined
To form that remarkable structure,
    it pleased him to call—his mind.

In another poem he wrote:

. . . While down the stream of Evolution
We drift, expecting no solution
But that of the survival of the fittest.
Till, in the twilight of the gods,
When earth and sun are frozen clods,
When, all its energy degraded,
Matter to aether shall have faded;
We, that is, all the work we've done,
As waves in aether, shall for ever run
In ever-widening spheres through heavens beyond the sun.

21

Newton, The Man

(From a MS. by the late LORD KEYNES)

Read by MR GEOFFREY KEYNES, M.D., F.R.C.S.

The MS. of this lecture was actually prepared by the late Lord Keynes some years ago for delivery to an audience at Trinity College. It was also read to a small private audience at the Royal Society Club in 1942.

Although he had intended to use this MS. as the basis of his lecture on this occasion, he would no doubt have made certain alterations. His draft as read by Mr Geoffrey Keynes must be regarded as unfinished.

It is with some diffidence that I try to speak to you in his own home of Newton *as he was himself.* I have long been a student of the records and had the intention to put my impressions into writing to be ready for Christmas Day 1942, the tercentenary of his birth. The war has deprived me both of leisure to treat adequately so great a theme and of opportunity to consult my library and my papers and to verify my impressions. So if the brief study which I shall lay before you to-day is more perfunctory than it should be, I hope you will excuse me.

One other preliminary matter. I believe that Newton was different from the conventional picture of him. But I do not believe he was less great. He was less ordinary, more extraordinary, than the nineteenth century cared to make him out. Geniuses *are* very peculiar. Let no one here suppose that my object to-day is to lessen, by describing, Cambridge's greatest son. I am trying rather to see him as his own friends and contemporaries saw him. And they without exception regarded him as one of the greatest of men.

In the eighteenth century and since, Newton came to be thought of as the first and greatest of the modern age of scientists, a rationalist, one who taught us to think on the lines of cold and untinctured reason.

I do not see him in this light. I do not think that any one who has pored over the contents of that box which he packed up when he finally left Cambridge in 1696 and which, though partly dispersed, have come down to us, can see him like that. Newton was not the first of the age of reason. He was the last of the magicians, the last of the Babylonians and Sumerians, the last great mind which looked out on the visible and intellectual world with the same eyes as those who began to build our intellectual inheritance rather less than 10,000 years ago. Isaac Newton, a posthumous child born with no father on Christmas Day, 1642, was the last wonder-child to whom the Magi could do sincere and appropriate homage.

Had there been time, I should have liked to read to you the contemporary record of the child Newton. For, though it is well known to his biographers, it has never been published *in extenso*, without comment, just as it stands. Here, indeed, is the makings of a legend of the young magician, a most joyous picture

27

of the opening mind of genius free from the uneasiness, the melancholy and nervous agitation of the young man and student.

For in vulgar modern terms Newton was profoundly neurotic of a not unfamiliar type, but—I should say from the records—a most extreme example. His deepest instincts were occult, esoteric, semantic—with profound shrinking from the world, a paralyzing fear of exposing his thoughts, his beliefs, his discoveries in all nakedness to the inspection and criticism of the world. 'Of the most fearful, cautious and suspicious temper that I ever knew', said Whiston, his successor in the Lucasian Chair. The too well-known conflicts and ignoble quarrels with Hooke, Flamsteed, Leibnitz are only too clear an evidence of this. Like all his type he was wholly aloof from women. He parted with and published nothing except under the extreme pressure of friends. Until the second phase of his life, he was a wrapt, consecrated solitary, pursuing his studies by intense introspection with a mental endurance perhaps never equalled.

I believe that the clue to his mind is to be found in his unusual powers of continuous concentrated introspection. A case can be made out, as it also can with Descartes, for regarding him as an accomplished experimentalist. Nothing can be more charming than the tales of his mechanical contrivances when he was a boy. There are his telescopes and his optical experiments. These were essential accomplishments, part of his unequalled all-round technique, but not, I am sure, his *peculiar* gift, especially amongst his contemporaries. His peculiar gift was the power of holding continuously in his mind a purely mental problem until he had seen straight through it. I fancy his pre-eminence is due to his muscles of intuition being the strongest and most enduring with which a man has ever been gifted. Anyone who has ever attempted pure scientific or philosophical thought knows how one can hold a problem momentarily in one's mind and apply all one's powers of concentration to piercing through it, and how it will dissolve and escape and you find that what you are surveying is a blank. I believe that Newton could hold a problem in his mind for hours and days and weeks until it surrendered to him its secret. Then being a supreme mathematical technician he could dress it up, how you will, for purposes of exposition, but it was his intuition which was pre-eminently extraordinary—'so happy in his conjectures', said de Morgan, 'as to seem to know more than he could possibly have any means of proving'. The proofs, for what they are worth, were, as I have said, dressed up afterwards—they were not the instrument of discovery.

There is the story of how he informed Halley of one of his most fundamental discoveries of planetary motion. 'Yes,' replied Halley, 'but how do you know that? Have you proved it?' Newton was taken aback—'Why, I've known it for years', he replied. 'If you'll give me a few days, I'll certainly find you a proof of it'—as in due course he did.

Again, there is some evidence that Newton in preparing the *Principia* was held up almost to the last moment by lack of proof that you could treat a solid

28

sphere as though all its mass was concentrated at the centre, and only hit on the proof a year before publication. But this was a truth which he had known for certain and had always assumed for many years.

Certainly there can be no doubt that the peculiar geometrical form in which the exposition of the *Principia* is dressed up bears no resemblance at all to the mental processes by which Newton actually arrived at his conclusions.

His experiments were always, I suspect, a means, not of discovery, but always of verifying what he knew already.

Why do I call him a magician? Because he looked on the whole universe and all that is in it *as a riddle*, as a secret which could be read by applying pure thought to certain evidence, certain mystic clues which God had laid about the world to allow a sort of philosopher's treasure hunt to the esoteric brotherhood. He believed that these clues were to be found partly in the evidence of the heavens and in the constitution of elements (and that is what gives the false suggestion of his being an experimental natural philosopher), but also partly in certain papers and traditions handed down by the brethren in an unbroken chain back to the original cryptic revelation in Babylonia. He regarded the universe as a cryptogram set by the Almighty—just as he himself wrapt the discovery of the calculus in a cryptogram when he communicated with Leibnitz. By pure thought, by concentration of mind, the riddle, he believed, would be revealed to the initiate.

He *did* read the riddle of the heavens. And he believed that by the same powers of his introspective imagination he would read the riddle of the Godhead, the riddle of past and future events divinely fore-ordained, the riddle of the elements and their constitution from an original undifferentiated first matter, the riddle of health and of immortality. All would be revealed to him if only he could persevere to the end, uninterrupted, by himself, no one coming into the room, reading, copying, testing—all by himself, no interruption for God's sake, no disclosure, no discordant breakings in or criticism, with fear and shrinking as he assailed these half-ordained, half-forbidden things, creeping back into the bosom of the Godhead as into his mother's womb. 'Voyaging through strange seas of thought *alone*', not as Charles Lamb 'a fellow who believed nothing unless it was as clear as the three sides of a triangle'.

And so he continued for some twenty-five years. In 1687, when he was forty-five years old, the *Principia* was published.

Here in Trinity it is right that I should give you an account of how he lived amongst you during these years of his greatest achievement. The east end of the Chapel projects farther eastwards than the Great Gate. In the second half of the seventeenth century there was a walled garden in the free space between Trinity Street and the building which joins the Great Gate to the Chapel. The south wall ran out from the turret of the Gate to a distance overlapping the Chapel by at least the width of the present pavement. Thus the garden was of

29

24

modest but reasonable size, as is well shown in Loggan's print of the College (reproduced here from *Cantabrigia Illustrata*, 1688). This was Newton's garden. He had the Fellow's set of rooms between the Porter's Lodge and the Chapel—that, I suppose, now occupied by Professor Broad. The garden was reached by a stairway which was attached to a veranda raised on wooden pillars projecting into the garden from the range of buildings. At the top of this stairway stood his telescope—not to be confused with the observatory erected on the top of the Great Gate during Newton's lifetime (but after he had left Cambridge) for the use of Roger Cotes and Newton's successor, Whiston. This wooden erection was, I think, demolished by Whewell in 1856 and replaced by the stone bay of Professor Broad's bedroom. At the Chapel end of the garden was a small two-storied building, also of wood, which was his elaboratory. When he decided to prepare the *Principia* for publication he engaged a young kinsman, Humphrey Newton, to act as his amanuensis (the MS. of the *Principia*, as it went to the press, is clearly in the hand of Humphrey). Humphrey remained with him for five years—from 1684 to 1689. When Newton died Humphrey's son-in-law Conduitt wrote to him for his reminiscences, and among the papers I have is Humphrey's reply.

During these twenty-five years of intense study mathematics and astronomy were only a part, and perhaps not the most absorbing, of his occupations. Our record of these is almost wholly confined to the papers which he kept and put in his box when he left Trinity for London.

Let me give some brief indications of their subject. They are enormously voluminous—I should say that upwards of 1,000,000 words in his handwriting still survive. They have, beyond doubt, no substantial value whatever except as a fascinating sidelight on the mind of our greatest genius.

Let me not exaggerate through reaction against the other Newton myth which has been so sedulously created for the last two hundred years. There was extreme method in his madness. All his unpublished works on esoteric and theological matters are marked by careful learning, accurate method and extreme sobriety of statement. They are just as *sane* as the *Principia*, if their whole matter and purpose were not magical. They were nearly all composed during the same twenty-five years of his mathematical studies. They fall into several groups.

Very early in life Newton abandoned orthodox belief in the Trinity. At this time the Socinians were an important Arian sect amongst intellectual circles. It may be that Newton fell under Socinian influences, but I think not. He was rather a Judaic monotheist of the school of Maimonides. He arrived at this conclusion, not on so-to-speak rational or sceptical grounds, but entirely on the interpretation of ancient authority. He was persuaded that the revealed documents give no support to the Trinitarian doctrines which were due to late falsifications. The revealed God was one God.

But this was a dreadful secret which Newton was at desperate pains to conceal

30

all his life. It was the reason why he refused Holy Orders, and therefore had to obtain a special dispensation to hold his Fellowship and Lucasian Chair and could not be Master of Trinity. Even the Toleration Act of 1689 excepted anti-Trinitarians. Some rumours there were, but not at the dangerous dates when he was a young Fellow of Trinity. In the main the secret died with him. But it was revealed in many writings in his big box. After his death Bishop Horsley was asked to inspect the box with a view to publication. He saw the contents with horror and slammed the lid. A hundred years later Sir David Brewster looked into the box. He covered up the traces with carefully selected extracts and some straight fibbing. His latest biographer, Mr More, has been more candid. Newton's extensive anti-Trinitarian pamphlets are, in my judgement, the most interesting of his unpublished papers. Apart from his more serious affirmation of belief, I have a completed pamphlet showing up what Newton thought of the extreme dishonesty and falsification of records for which St Athanasius was responsible, in particular for his putting about the false calumny that Arius died in a privy. The victory of the Trinitarians in England in the latter half of the seventeenth century was not only as complete, but also as extraordinary, as St Athanasius's original triumph. There is good reason for thinking that Locke was a Unitarian. I have seen it argued that Milton was. It is a blot on Newton's record that he did not murmur a word when Whiston, his successor in the Lucasian Chair, was thrown out of his professorship and out of the University for publicly avowing opinions which Newton himself had secretly held for upwards of fifty years past.

That he held this heresy was a further aggravation of his silence and secrecy and inwardness of disposition.

Another large section is concerned with all branches of apocalyptic writings from which he sought to deduce the secret truths of the Universe—the measurements of Solomon's Temple, the Book of David, the Book of Revelations, an enormous volume of work of which some part was published in his later days. Along with this are hundreds of pages of Church History and the like, designed to discover the truth of tradition.

A large section, judging by the handwriting amongst the earliest, relates to alchemy—transmutation, the philosopher's stone, the elixir of life. The scope and character of these papers have been hushed up, or at least minimized, by nearly all those who have inspected them. About 1650 there was a considerable group in London, round the publisher Cooper, who during the next twenty years revived interest not only in the English alchemists of the fifteenth century, but also in translations of the medieval and post-medieval alchemists.

There is an unusual number of manuscripts of the early English alchemists in the libraries of Cambridge. It may be that there was some continuous esoteric tradition within the University which sprang into activity again in the twenty years from 1650 to 1670. At any rate, Newton was clearly an unbridled addict.

31

It is this with which he was occupied 'about 6 weeks at spring and 6 at the fall when the fire in the elaboratory scarcely went out' at the very years when he was composing the *Principia*—and about this he told Humphrey Newton not a word. Moreover, he was almost entirely concerned, not in serious experiment, but in trying to read the riddle of tradition, to find meaning in cryptic verses, to imitate the alleged but largely imaginary experiments of the initiates of past centuries. Newton has left behind him a vast mass of records of these studies. I believe that the greater part are translations and copies made by him of existing books and manuscripts. But there are also extensive records of experiments. I have glanced through a great quantity of this—at least 100,000 words, I should say. It is utterly impossible to deny that it is wholly magical and wholly devoid of scientific value; and also impossible not to admit that Newton devoted years of work to it. Some time it might be interesting, but not useful, for some student better equipped and more idle than I to work out Newton's exact relationship to the tradition and MSS. of his time.

In these mixed and extraordinary studies, with one foot in the Middle Ages and one foot treading a path for modern science, Newton spent the first phase of his life, the period of life in Trinity when he did all his real work. Now let me pass to the second phase.

After the publication of the *Principia* there is a complete change in his habit and way of life. I believe that his friends, above all Halifax, came to the conclusion that he must be rooted out of the life he was leading at Trinity which must soon lead to decay of mind and health. Broadly speaking, of his own motion or under persuasion, he abandons his studies. He takes up University business, represents the University in Parliament; his friends are busy trying to get a dignified and remunerative job for him—the Provostship of King's, the Mastership of Charterhouse, the Controllership of the Mint.

Newton could not be Master of Trinity because he was a Unitarian and so not in Holy Orders. He was rejected as Provost of King's for the more prosaic reason that he was not an Etonian. Newton took this rejection very ill and prepared a long legalistic brief, which I possess, giving reasons why it was not unlawful for him to be accepted as Provost. But, as ill-luck had it, Newton's nomination for the Provostship came at the moment when King's had decided to fight against the right of Crown nomination, a struggle in which the College was successful.

Newton was well qualified for any of these offices. It must not be inferred from his introspection, his absent-mindedness, his secrecy and his solitude that he lacked aptitude for affairs when he chose to exercise it. There are many records to prove his very great capacity. Read, for example, his correspondence with Dr Covell, the Vice-Chancellor when, as the University's representative in Parliament, he had to deal with the delicate question of the oaths after the revolution of 1688. With Pepys and Lowndes he became one of the greatest and

32

most efficient of our civil servants. He was a very successful investor of funds, surmounting the crisis of the South Sea Bubble, and died a rich man. He possessed in exceptional degree almost every kind of intellectual aptitude—lawyer, historian, theologian, not less than mathematician, physicist, astronomer.

And when the turn of his life came and he put his books of magic back into the box, it was easy for him to drop the seventeenth century behind him and to evolve into the eighteenth-century figure which is the traditional Newton.

Nevertheless, the move on the part of his friends to change his life came almost too late. In 1689 his mother, to whom he was deeply attached, died. Somewhere about his fiftieth birthday on Christmas Day 1692, he suffered what we should now term a severe nervous breakdown. Melancholia, sleeplessness, fears of persecution—he writes to Pepys and to Locke and no doubt to others letters which lead them to think that his mind is deranged. He lost, in his own words, the 'former consistency of his mind'. He never again concentrated after the old fashion or did any fresh work. The breakdown probably lasted nearly two years, and from it emerged, slightly 'gaga', but still, no doubt, with one of the most powerful minds of England, the Sir Isaac Newton of tradition.

In 1696 his friends were finally successful in digging him out of Cambridge, and for more than another twenty years he reigned in London as the most famous man of his age, of Europe, and—as his powers gradually waned and his affability increased—perhaps of all time, so it seemed to his contemporaries.

He set up house with his niece Catharine Barton, who was beyond reasonable doubt the mistress of his old and loyal friend Charles Montague, Earl of Halifax and Chancellor of the Exchequer, who had been one of Newton's intimate friends when he was an undergraduate at Trinity. Catharine was reputed to be one of the most brilliant and charming women in the London of Congreve, Swift and Pope. She is celebrated, not least for the broadness of her stories, in Swift's *Journal to Stella*. Newton puts on rather too much weight for his moderate height. 'When he rode in his coach one arm would be out of his coach on one side and the other on the other.' His pink face, beneath a mass of snow-white hair, which 'when his peruke was off was a venerable sight', is increasingly both benevolent and majestic. One night in Trinity after Hall he is knighted by Queen Anne. For nearly twenty-four years he reigns as President of the Royal Society. He becomes one of the principal sights of London for all visiting intellectual foreigners, whom he entertains handsomely. He liked to have clever young men about him to edit new editions of the *Principia*—and sometimes merely plausible ones as in the case of Facio de Duillier.

Magic was quite forgotten. He has become the Sage and Monarch of the Age of Reason. The Sir Isaac Newton of orthodox tradition—the eighteenth-century Sir Isaac, so remote from the child magician born in the first half of the seventeenth century—was being built up. Voltaire returning from his trip to London

28

was able to report of Sir Isaac—''twas his peculiar felicity, not only to be born in a country of liberty, but in an Age when all scholastic impertinences were banished from the World. Reason alone was cultivated and Mankind cou'd only be his Pupil, not his Enemy.' Newton, whose secret heresies and scholastic superstitions it had been the study of a lifetime to conceal!

But he never concentrated; never recovered 'the former consistency of his mind'. 'He spoke very little in company.' 'He had something rather languid in his look and manner.'

And he looked very seldom, I expect, into the chest where, when he left Cambridge, he had packed all the evidences of what had occupied and so absorbed his intense and flaming spirit in his rooms and his garden and his elaboratory between the Great Gate and Chapel.

But he did not destroy them. They remained in the box to shock profoundly any eighteenth- or nineteenth-century prying eyes. They became the possession of Catharine Barton and then of her daughter, the Countess of Portsmouth. So Newton's chest, with many hundreds of thousands of words of his unpublished writings, came to contain the 'Portsmouth Papers'.

In 1888 the mathematical portion was given to the University Library at Cambridge. They have been indexed, but they have never been edited. The rest, a very large collection, were dispersed in the auction room in 1936 by Catharine Barton's descendant, the present Lord Lymington. Disturbed by this impiety, I managed gradually to reassemble about half of them, including nearly the whole of the biographical portion, that is, the 'Conduitt Papers', in order to bring them to Cambridge which I hope they will never leave. The greater part of the rest were snatched out of my reach by a syndicate which hoped to sell them at a high price, probably in America, on the occasion of the recent tercentenary.

As one broods over these queer collections, it seems easier to understand— with an understanding which is not, I hope, distorted in the other direction— this strange spirit, who was tempted by the Devil to believe at the time when within these walls he was solving so much, that he could reach *all* the secrets of God and Nature by the pure power of mind—Copernicus and Faustus in one.

34

29

Einstein: Early Years

Einstein hated most things that other men hold dear. 'Comfort and happiness,' he declared in later life, 'have never appeared to me as a goal. I call these ethical bases the ideals of the swineherd. . . . The commonplace goals of human endeavour – possessions, outward success and luxury have always seemed to me despicable, since early youth.' He was essentially a lone wolf. He never took part in any student gatherings, but although he rejected the churches he had a Spinoza-like belief in a cosmic religious force. He regarded this as an eternal spiritual being that communicates small details of itself to our weak and inadequate minds. As he once declared, 'This deep intuitive conviction of the existence of a higher power of thought which manifests itself in the inscrutable universe represents the content of my definition of God.' In other words, he had no more use for the shallow materialism that is the most widely accepted philosophy of scientists and others today than for the authoritarian views of the churches that were once so powerful. Of all religious bodies, the one that he felt most sympathy for was the pacificist Society of Friends, the Quakers. And in this connection it is perhaps not altogether irrelevant to mention that the first eminent scientist to expound Einstein's theory of relativity in this country was, in fact, a Quaker, the late Sir Arthur Eddington.

The outlook for Einstein after graduating was bleak. An aunt in Genoa who had made him a monthly allowance ceased to pay it. Soon afterwards his father died and his mother had to work as a housekeeper. As one of his biographers has said, 'That he managed to maintain himself financially in Zürich

6

until the autumn of 1901 can be attributed to Einstein's stoical indifference to material things.' He did some calculations for the director of the Federal Observatory and then taught for a few months in a technical school in Winterthur while the resident teacher of mathematics was doing his military service. Meanwhile he became a Swiss citizen of Zürich, and as he never repudiated it he retained this citizenship until his death. He was never called for military service, however, because he failed his medical on account of flat feet and varicose veins, a verdict that irritated him!

At last, in 1902, his friend Marcel Grossmann introduced him to the director of the Swiss Patent Office in Berne, who decided to appoint him to his staff, in June of that year, at a salary that enabled him to live quite comfortably. Soon afterwards he married Mileva. Although the marriage seems to have been reasonably happy at first, it eventually broke up. They had two sons.

Einstein's move to Berne was a turning-point in his life. Although he had had no previous experience with technical inventions, he found the work in the Patent Office interesting. It was his duty to put applications into a clear form and to determine the basic idea from the often vaguely worded descriptions of the inventors. It may well have been this training that developed his remarkable faculty for seeing to the heart of a problem and quickly realizing the consequences of any hypothesis. Moreover, the work left him with ample time to pursue his own ideas. Indeed, it would seem in many ways to have been the ideal post for him at this stage of his career.

Einstein's first scientific paper appeared in 1901. It was on capillary attraction. Among other early papers were two, published in 1902 and 1903, on the statistical foundations of thermodynamics, a theory that had made a profound impression on him. His work amounted to a new derivation of the main features of statistical mechanics by a similar method to that developed a year or two earlier by the American mathe-

7

matical physicist Willard Gibbs, although Einstein was unaware of this. Einstein's approach was less abstract than that of Gibbs and he went on to consider a practical application of the greatest significance.

At this time the reality of molecules and of the kinetic theory of matter, according to which the temperature of a body is due to the thermal agitation of its constituent molecules, were still under dispute. Einstein discovered that this thermal agitation can produce a visible and measurable effect on particles suspended in a solution. Such an effect had, in fact, been discovered by the Scottish botanist Robert Brown as long ago as 1827, when he observed that pollen grains suspended in water became dispersed in a great number of small particles which were in constant motion, moving in irregular zigzags even in the absence of currents and other external disturbances.

The fundamental paper in which Einstein showed that the Brownian motion could be used as direct evidence for the existence of molecules and the correctness of the kinetic theory of heat was published in 1905. He argued that, although the velocity of a suspended particle due to the impacts of the molecules of the liquid on it is unobservable, the effect of a succession of irregular displacements can be detected with the aid of a microscope. The actual observations were made later by the French physicist Jean Perrin and completely confirmed Einstein's theory. This pioneer analysis of a phenomenon of thermal fluctuations has been the inspiration for many later investigations of similar effects. Moreover, the phenomenon of Brownian motion has come to be regarded as one of the best 'direct' proofs of the existence of molecules. Einstein's researches on this subject also played their part in helping to convince physicists of the importance of probability in relation to natural laws. Nevertheless, Einstein himself seems always to have believed that the ultimate laws are essentially causal and deterministic and that it is only our inability to

8

deal with large numbers of particles in any other way that compels us to use statistical methods.

The year 1905 was Einstein's *annus mirabilis*. Because of his contributions, volume 17 of the *Annalen der Physik* of that year is now regarded as one of the most remarkable volumes of scientific literature ever published. A copy was recently offered for sale by a London dealer at no less than £550! It contains three papers by Einstein. Each is on a different subject and is a masterpiece. The paper on the Brownian motion was the second of these. It was preceded by Einstein's first contribution to quantum physics. Like the third paper, which was on relativity, it was concerned with the behaviour of light.

One of the problems that puzzled physicists at the end of the nineteenth century concerned the light radiated from a hot body. As the temperature rises, the body changes in colour from a dull glaring red to a brighter orange and then to a blinding bluish white. Attempts to explain this change in the quality of light with temperature had failed to show why at a given temperature there is no perceptible radiation above a certain frequency of vibration corresponding to light of a definite colour. It appeared that it must be difficult to emit light of very high frequency, but no one knew why. To resolve the problem Max Planck assumed, in 1900, that the energy of radiation emitted with a given frequency v by an oscillating atom cannot be less than a definite minimum amount hv and must increase by units of this amount, where h is a new universal constant of nature, now known as Planck's constant. Thus light of a given frequency, or colour, can only be emitted and absorbed in discrete packets of energy, or quanta, and not in purely arbitrary amounts. With this assumption, Planck derived results in the theory of radiation that agreed with observation, and even succeeded in calculating the value of his new constant h.

Planck's hypothesis was slow to make its effect felt on physics, because it seemed difficult to reconcile with the generally

9

accepted wave-like character of radiation. Planck still believed that light and all other forms of electromagnetic radiation were continuously divisible and that the quantum effects concerned only their interaction with matter. This contrast between the continuous nature of radiation and the discontinuous character of its emission and absorption by atoms puzzled Einstein. He came to the conclusion that radiation itself must have a corpuscular structure and that it is actually *composed* of Planck's quanta. He showed that this hypothesis provided a simple explanation of a curious physical phenomenon, known as the photoelectric effect, that had been a great puzzle to physicists for some years. It had been found that when light shines on metallic bodies electrons are emitted if, and only if, the frequency of the light exceeds a certain threshold value depending on the particular metal. Moreover, the speed with which the electrons are emitted does not depend on the intensity of the light, but only on its frequency, or colour.

Einstein showed that these observations could be explained if the energy in a beam of light of frequency ν were concentrated in small packets, each of energy $h\nu$ where h is Planck's constant. These packets of radiation are now called photons. For an electron to be ejected a certain amount of energy must be expended, depending on the nature of the metal. Consequently, if a bombarding photon is to liberate an electron from the metal, it must have sufficient energy and hence its frequency must be above a certain threshold value. Moreover, the speed with which the electron is ejected will depend on the excess energy of the incident photon, and this in turn will depend on the amount by which its frequency exceeds the threshold value. Einstein's simple formula for the speed of the emitted electrons was later confirmed experimentally and a value obtained for h in good agreement with Planck's original estimate.

In introducing his quantum hypothesis, Planck seems to have thought that he was making only a minor adjustment to

10

the classical laws of physics. Einstein believed that a much more radical point of view was necessary and that the whole structure of mechanistic physics based on Newton's laws of motion was at stake. This belief found further expression in the third of his great papers of 1905. It originated in another problem concerning the behaviour of light and gave rise to what is now known as the special theory of relativity.

It happens that we have detailed evidence concerning the way in which Einstein gradually came to develop this theory. For, in 1916, the distinguished psychologist Max Wertheimer had several long conversations with Einstein on this question. He afterwards presented a fascinating account in one of the chapters of his book *Productive Thinking*. Already, at the age of sixteen, Einstein had been puzzled by the following paradox. According to generally accepted ideas, a beam of light travels in empty space with a finite velocity of approximately 300,000 kilometres a second. Einstein tried to imagine what he would observe if he were to travel through space with the same velocity as such a beam. According to the usual idea of relative motion, it would seem that the beam of light would then appear as a spatially oscillating electromagnetic field at rest. But such a concept was unknown to physics and at variance with Maxwell's theory. Einstein began to suspect that the laws of physics, including those concerning the propagation of light, must remain the same for all observers however fast they move relative to one another.

When Wertheimer asked Einstein if already at this time he had some idea of the invariance of the velocity of light for all observers in uniform relative motion, Einstein replied, 'No, it was just a curiosity. That the velocity of light could differ depending on the movement of the observer was somehow characterized by doubt. Later developments increased that doubt.'

Nevertheless, as he told Wertheimer, it was only after years of thought that he finally felt compelled to regard the velocity

11

of light as a fundamental invariant independent of the motion of the observer, for this idea conflicted with traditional views concerning the measurement of motion. How, then, is motion to be measured? Einstein realized that it must depend on the measurement of time. 'Do I see clearly', he asked himself, 'the relation, the inner connection between the measurement of time and that of movement?' It occurred to him that time measurement depends on the idea of simultaneity. Suddenly he was struck by the fact that, although this idea was perfectly clear when two events occur in the same place, it was not equally clear for events in different places. This was the crucial stage in his thinking. For he saw that he had discovered a great gap in the classical treatment of time. It took him roughly ten years to arrive at this point, but from the moment when he came to question the traditional idea of time, only five weeks were needed to write his paper, although he was working all day at the Patent Office.

The critical reasoning that led Einstein to abandon the classical concept of world-wide simultaneity was stimulated by his interest in philosophy. Soon after his move to Berne in 1902 he had made the acquaintance of a Rumanian named Maurice Solovine, who was studying both physics and philosophy, and of a Swiss student named Conrad Habicht. In the evenings they often read together and discussed classics of philosophy by Plato, Kant, Mill, Poincaré, and others.

According to Einstein's own account, the philosophers who helped him most to develop his critical powers were David Hume and Ernst Mach. Hume influenced Einstein by his penetrating criticism of traditional common-sense assumptions and dogmas. Mach's influence was more direct and at the same time more complex. For Einstein was not at all in sympathy with Mach's general philosophy of science based on the doctrine that the laws of physics are only summaries of experimental results. Instead, Einstein believed that these laws also involve factors contributed by the human mind. Nevertheless,

12

Mach influenced Einstein by his criticism of Newton's ideas concerning space and time and also by his critical examination of Newtonian mechanics.

While Einstein was gradually being led to question the classical conception of time, he was also becoming increasingly sceptical of the mechanistic idea that electromagnetic waves in empty space must be regarded as oscillations in a peculiar universal medium called the ether. Indeed, the properties of this medium seemed to defy mechanical explanation.

On the one hand, it seemed clear that the Earth does not drag the ether along with it in its motion around the Sun. This had been shown by Bradley's discovery that the apparent directions of the stars exhibit small annual changes due to the cyclical change in the Earth's direction of motion during the course of the year. For, if the ether in which the light waves from the stars undulate were dragged along by the Earth, no such effect would be observed. Consequently, the Earth must move through the ether.

If, however, the Earth moves through the ether, then it should be possible in principle to determine its velocity relative to the ether by measuring the velocity of light relative to the Earth in different directions. This had been the object of the famous Michelson–Morley experiment of 1887. Various attempts were made to explain its failure, despite the adequacy of the apparatus to detect the effect of a velocity through the ether considerably less than that with which the Earth was known to move around the Sun. Although Einstein's train of thought seems, in fact, to have been but little influenced by this particular experiment, when he eventually heard of it he immediately realized that the trouble was due to introducing the idea of the ether as the medium of the transmission of light. Just as Mach had rejected Newton's concept of absolute space, so Einstein discarded the luminiferous ether and with it the mechanical conception of optical phenomena. In his view, Michelson and Morley obtained a null result because,

13

37

despite the Earth's motion, their apparatus must have functioned exactly as if it had been at rest throughout the experiment.

A similar state of affairs had long been familiar in mechanics. It was known that mechanical forces produce the same effects on bodies in uniform motion as on bodies at rest. For example, any mechanical experiment performed on board a ship sailing steadily in a straight line yields the same result as a similar experiment carried out on land. Newton's laws of motion had therefore been formulated so as to be the same for all frames of reference in uniform relative motion, including those at relative rest. On this basis, it is impossible, even in principle, to devise a mechanical experiment to measure an absolute velocity.

In 1904 the great French mathematician Henri Poincaré introduced a general physical law to cover both this situation in mechanics and the result of the Michelson-Morley experiment in optics. He called it the 'principle of relativity'. According to this principle, now known as the 'principle of special relativity', the laws of physics are the same for all observers at rest or moving uniformly in straight lines. Nevertheless, although Poincaré's principle implies that, as far as experimental physics is concerned, all uniform motion is relative and no absolute uniform motion is detectable, he did not go on to construct a theory of relativity, as Einstein did shortly afterwards. For, unlike Einstein, Poincaré failed to realize that his principle made the ether concept unnecessary and indeed was not strictly compatible with it. Instead, he continued to believe that the ether must be retained as a mechanical basis for the transmission of light.

The true creator of the special theory of relativity was therefore Einstein in his paper of 1905, in which he accepted the principle of relativity unconditionally as a fundamental general law of physics. As we have already seen, he had been feeling his way towards this principle on what we may call

14

'philosophical' grounds for many years. He did not attempt to account for it in terms of other physical hypotheses, nor did he appeal to the Michelson-Morley experiment to justify it, but he regarded it as a more suitable starting-point for the general study of physical phenomena than Newton's laws. It was this, more than anything else, that made Einstein's theory so difficult for older physicists to accept. Instead of reducing optics to mechanics, Einstein based both optical and mechanical laws on the same fundamental principle of relativity.

In accordance with this principle, light should have the same properties for all observers and frames of reference in uniform relative motion. In particular, as Einstein was forced to conclude, its velocity in empty space *must* be the same for all of them, despite the fact that this condition clashes with the common-sense assumption of world-wide simultaneity. Einstein therefore abandoned this assumption and explored instead the consequences of using the invariance of the velocity of light as a means of comparing the time readings of clocks in uniform relative motion in different places. He found that, on this basis, different observers would, in general, assign different times to the same event and that a moving clock would appear to run slow compared with an identical clock at rest with respect to the observer. Although these effects are small, except when the relative motion is nearly that of light, they seemed paradoxical and were received with incredulity by many educated people, including most philosophers.

Einstein showed that his principle could be used to derive from the classical laws of physics, assumed to be valid for velocities that are small compared with the velocity of light, laws that are valid for all velocities. In particular, he found that Newton's laws of motion, formerly regarded as the foundation of physics, had to be modified for rapidly moving bodies. Thus the inertial mass of a body, formerly supposed

15

to be independent of motion, was now found to increase indefinitely the nearer its velocity approaches that of light. Consequently, a given force acting on the body will produce smaller and smaller changes in its velocity the faster it moves. As a result no particle of matter can ever attain the speed of light. In this way, Einstein solved his ten-year-old problem concerning the observer who moves with the same velocity as a beam of light. This motion is physically impossible.

The dependence of mass on velocity led Einstein to a remarkable unification of concepts. He came to the conclusion that mass and energy are intimately associated. Corresponding to any increase in the energy-content of a body, there is an equivalent increase in its mass. In a short paper, published later in 1905, Einstein similarly showed that, if a body gives off energy E in the form of radiation, its mass is diminished by an amount M, where $E = Mc^2$.

This unification of mass and energy was one of the most important consequences of Einstein's special theory of relativity. It implied that matter can be regarded as highly concentrated energy. This hypothesis has received remarkable confirmation in nuclear physics and has resolved the problem of the origin of the Sun's radiation. The enormous release of energy in nuclear reactions is due to the conversion of a small quantity of mass into its equivalent large amount of liberated energy.

To Einstein the main value of his discovery concerning mass and energy did not lie in its practical applications. The vital point was that he had come upon it as a consequence of the relativity principle. He believed that this principle was applicable to all natural phenomena and that it could unite the laws of nature. However, before he could successfully apply it to gravitation, he had to generalize the principle to cover accelerated motion, and this problem was to be his main preoccupation for the next ten years.

Although it was by this later work on general relativity

16

that Einstein was destined to attain world-wide popular fame, his professional standing among physicists can be traced back to the great papers of 1905. For it was in these papers that he first revealed the essential limitations of classical theories and laid the foundations of twentieth-century theoretical physics.

Today there must be very few people who can remember Albert Einstein as a young man in Berne. Professor André Mercier is a distinguished scientist living in that city who is head of the department of theoretical physics in the University of Berne and also Secretary-General of the International Committee on General Relativity and Gravitation. When he was last in London I asked him to tell me something about Einstein's early life in Switzerland.

Mercier: The fact that the young Albert was from his youth brought up in Switzerland certainly had a decisive influence on him. The simple and democratic way of Swiss life was in those days very different from that of neighbouring countries. College years spent in the mainly Protestant environment revealed to Einstein that he did not easily conform to Jewish traditional orthodoxy so that, although in some circles it has become fashionable to insist upon Einstein's Jewishness, he himself never felt, except possibly in his later years, that he primarily belonged to that creed and tradition. He was much too cosmopolitan for that.

Whitrow: Was not Einstein a Swiss citizen?

Mercier: Well, that is an interesting question. When he came to our country as a boy he was by nationality a German and remained so until he became of age; then he expressly applied for Swiss citizenship. Although it is notoriously difficult to acquire Swiss citizenship, it was granted to him. Later when he went to Berlin he was again made a German citizen and many years later, after he had settled at Princeton, New Jersey, American citizenship was conferred upon him by an act of Congress, but these successive nationalities were bestowed upon him almost like honorary degrees. Nevertheless,

17

he retained his Swiss citizenship until the end of his life. In virtue of this, he had a traditionally international neutral status, and he was certainly vividly aware of its significance. In this connection it may be mentioned that the only diploma he had on the walls of his office in Princeton was that of an honorary member of the Berne Society of Sciences. The years Einstein spent in Berne, where he wrote his first revolutionary scientific papers and where he was virtually unknown except to a handful of dear friends, were probably the happiest of his life.

Whitrow: You may have known some of his friends in Berne. Could you describe one or two of them?

Mercier: Well, it is not so easy and I shall not dwell upon his first marriage, which brought him to a tiny flat in the picturesque medieval part of the city. One friend had an important influence on Einstein's early scientific career. This was Michele Besso, a clever but rather queer engineer from Ticino, who was a colleague in the Swiss Patent Office in Berne. A man with Latin manners – he spoke better French than German and could talk endlessly – he was capable of showing great patience in discussion. He came to be called the 'sounding-board' for his friend Albert.

Although there is no doubt that Einstein conceived his theories purely by himself, the form in which he chose to communicate them to the scientific world was considerably influenced by the endless discussions he had with Besso, and this shows that, even in the case of the most original scientific geniuses, discussion with others is invaluable.

Whitrow: As I have already mentioned, Einstein had two sons by his first wife, Mileva. The first, Hans, was born in 1904, and the second, Edward, in 1910. Edward's health broke down while he was still a boy and later he developed a mild form of schizophrenia. He died recently in Switzerland. Hans emigrated to the United States before the Second World War and is now a professor of hydraulics in the University of

18

California at Berkeley. Recently he was interviewed by Bernard Mayes on his youthful recollections of his father.

H. A. Einstein: One of my earliest memories goes back to when I must have been around three years old, maybe four, when my father made me a little cable car out of matchboxes. I remember that that was one of the nicest toys I had at the time and it worked. Out of just a little string and matchboxes and so on, he could make the most beautiful things. As a matter of fact, he always liked to improvise things of that sort, just as he would also like to improvise in his work in a way: for instance, when he had to give a talk he never knew ahead of time exactly what he was going to say. It would depend on the impression he got from the audience in which way he would express himself and into how much detail he would go. And so this improvisation was a very important part of his character and of his way of working. In other respects he had a character more like that of an artist than of a scientist as we usually think of them. For instance, the highest praise for a good theory or a good piece of work was not that it was correct nor that it was exact but that it was beautiful.

Mayes: What was he like as a father around the house? Did he take any interest in home life?

H. A. Einstein: Oh yes. I remember very well a time when he was still splitting wood for the stove and carrying coals up to heat in winter, because that was the way it was done at the beginning of the century. And, although he was not particularly clever with his hands to do more delicate things, he was always willing to help.

Mayes: And your mother, how did she treat his rising fame?

H. A. Einstein: She was proud of him, but that is as far as it went. It was very hard to understand, because she originally had studied with him and had been a scientist herself. But, somehow or other, with the marriage she gave up practically all of her ambitions in that direction.

19

43

Mayes: Was this the reason for their separation eventually?

H. A. Einstein: I doubt it. Why the separation came is something that was never quite clear to me. Trying to reconstruct it all afterwards, particularly from some of his own utterances, it seems that he had the impression that the family was taking a bit too much of his time, and that he had the duty to concentrate completely on his work. Personally, I do not believe that he ever achieved that, because in the family he actually had more time than when he had to look after himself and fight all the outside world alone.

Mayes: And how did your mother take the separation?

H. A. Einstein: Very hard.

Mayes: And you yourself, did it affect you?

H. A. Einstein: Yes. For quite a while, for a number of years, the separation was just a *de facto* separation and not formalized in any way. And that was probably the worst time, because then nobody knew what the future would bring – whether this was just a temporary condition or whether the marriage would end finally. It was particularly hard because it was during the war. Naturally, when everybody knew what was going to happen, then one could adjust to it.

Mayes: You say that at first the separation was *de facto*. When was that?

H. A. Einstein: In 1914.

Mayes: Did your mother communicate with your father subsequently?

H. A. Einstein: Oh yes.

Mayes: Did he feel rather cold towards her?

H. A. Einstein: I have the impression that he never felt what you call cold towards her. I mean, they may have had certain differences that I do not know of and do not understand and could not discuss, but they always communicated in a very personal and in a rather warm way.

Mayes: What about your birthday, did he remember that?

H. A. Einstein: No, he never did.

20

Mayes: He never remembered your birthday?

H. A. Einstein: He never did. For instance, when I was fifty years old I got a very nice letter from him, but the very first sentence was, 'Unfortunately, I have to admit that I didn't think about it, but your wife wrote me.'

Mayes: Tell me about your early days when you were a happy family together. Did you go on vacations together? Did he take that much interest in you?

H. A. Einstein: Yes. I do not remember any very extended vacation, but we often took small trips together and sometimes longer ones. In those days we were a happy family and there was nothing to indicate the separation to come.

Mayes: What sort of places did you go to?

H. A. Einstein: He was very fond of nature. He did not care for large, impressive mountains, but he liked surroundings that were gentle and colourful and gave one lightness of spirit. He needed this kind of relaxation from his intense work.

Mayes: And he was able to play with his children?

H. A. Einstein: We would play together with my toys, but he was also trying to educate us in a wider sense than the education one gets in school. He often told me that one of the most important things in his life was music. Whenever he felt that he had come to the end of the road or into a difficult situation in his work he would take refuge in music and that would usually resolve all his difficulties.

Mayes: Did he take part in disciplining you if you did things that merited it?

H. A. Einstein: Oh yes, when he felt it was necessary. And every once in a while he felt it was. I think I was quite a rascal.

Mayes: What did he do?

H. A. Einstein: Oh, he beat me up, just like anyone else would do.

Mayes: What did he beat you with? With a cudgel or something?

21

H. A. Einstein: Oh, I don't remember, but he did anyway.

Mayes: When he realized that he had discovered something tremendous, how did he react to this?

H. A. Einstein: Oh, just like a child. He was happy. He was walking around and telling everybody and all of a sudden would start whistling, you know. But he was always extremely careful about his findings. That means he would never accept anything until he had tested it all the way through in all directions.

Mayes: How was your relationship after you were grown up and worked independently?

H. A. Einstein: Very cordial and friendly. With both of us rather busy, the occasions of getting together were not frequent, but whenever we met we mutually reported on all the interesting developments in our fields and in our work. It may be somewhat astonishing that a theoretically-oriented mind as that of Albert Einstein would be interested in technical matters. But he thoroughly enjoyed learning about clever inventions and solutions, as he had always loved to solve certain types of puzzles. Maybe both, inventions and puzzles, reminded him of the happy, carefree and successful days at the patent office in Berne, the days before the first world war and all that followed.

Whitrow: We have seen that, as a young man in Berne, Einstein often discussed famous philosophical works with his friends. Few philosophers, however, have made any serious attempt to study Einstein's work. One who has is Bertrand Russell. Recently we asked him if he thought that Einstein was a scientist whom philosophers should study.

Russell: Einstein's stature as a scientist was, and remains, very high. He removed the mystery from gravitation which everybody since Newton had accepted with a reluctant feeling that it was unintelligible. If Einstein's reputation has appeared to diminish, that is only because recent work in physics has been mainly concerned with quantum theory. I do not think

22

that the work of our century in either relativity or quantum theory has had any very good influence upon philosophy, but I regard this as the fault of the philosophers, who, for the most part, have not thought it necessary to master modern physics. I hope that an increasing proportion of philosophers will, as time goes on, become aware that ignorance of physics condemns any philosophy to futility.

Whitrow: I am afraid that Bertrand Russell's criticism is still true of all too many philosophers. There are, however, some exceptions; one of them is Sir Karl Popper. I asked him to tell us something about the influence that Einstein has had on his own philosophy.

Popper: Einstein's influence on my thinking has been immense. I might even say that what I have done is mainly to make explicit certain points which are implicit in the work of Einstein. I will try to sum up in four points what I have learned from Einstein directly and indirectly:

(1) Even the best-established scientific theory, such as Newton's theory of gravitation or Fresnel's theory of light, may be overthrown, or corrected, as Einstein has shown. Consequently, even the best-established scientific theory always remains a hypothesis, a conjecture.

(2) The recognition of this fact can be and should be of outstanding importance for one's own scientific work. It certainly was so for Einstein's work. He was never satisfied with any of the theories he proposed. He always tried to probe into the weak spots in order to find their limitations. And he did find them: again and again did he criticize his own work in his papers. For example, he began his famous paper of 1915 in which he first proposed the field equations for gravitation with the statement that some of his previous papers were utterly mistaken; and similarly he wrote in 1918, while replying to some criticism, that he had so far failed to distinguish between two different principles, and that his failure had led to confusion.

23

(3) This attitude, which may be called the critical attitude, is characteristic of the best scientific activity.

(4) With Einstein's work it became very clear that this attitude of criticism was in science something fundamentally different from what philosophers consider and describe as the 'critical attitude', or the 'sceptical attitude', or the 'attitude of doubt'.

Whitrow: Could you elaborate the difference between the critical attitude of scientists and of philosophers?

Popper: Yes. When philosophers speak of criticism they have in mind something like this. A philosopher, say Mr Adam, proposes a philosophical theory and tries to give arguments which would prove it or justify the claim that it is a true theory. Thereupon another philosopher, Mr Baker, analyses Mr Adam's proof and shows that it is invalid. Mr Baker's destructive analysis of the claims of Mr Adam to have established his theory is what philosophers usually have in mind when they speak of criticism. Or to put it another way: philosophers usually mean by criticism an analysis that aims at showing the invalidity of some arguments which have been offered in justification of the claim that a certain theory is true.

Now, it seems to have been rarely recognized that criticism in science has a very different aim and character. It is not an attack upon the proof or the justification of a scientific theory, but an attack upon the theory itself; not an attack on the claim that the theory can be *shown* to be true, but an attack on what the theory itself tells us – on its content or its *consequences*. This is so because, especially since Einstein, scientists do not seriously hold that their theories can be true or 'verified'. Nowadays they will hardly claim more than that one theory can explain more facts than other known theories, or the same facts better; that it can be tested at least as well as these other theories or even better; and that it stands up to these tests at least as well as these other theories.

This attitude became particularly clear in the case of

24

Einstein's criticism of Newton. Newton, in fact, had claimed that his laws of motion were not conjectural but true descriptions (if not explanations) of the facts, and that they were established by induction. But Einstein, who was a great admirer of Newton, did not criticize this mistaken claim. He did something more important; he revolutionized physics by producing an alternative to Newton's theory which not only passed all the tests which Newton's theory had passed, but also certain tests which it had failed to pass, and a few further tests which altogether went beyond the range of application of Newton's theory of gravitation. Nevertheless, Einstein regarded his own theory of gravitation merely as a step towards a better theory. Thus he wrote about his own field equations of gravitation that, as a matter of course, he never thought for a moment that his formulation of the field equations was more than a makeshift, designed to present provisionally the general principle of relativity in a concise form. And at the end of his last work, published in 1955, when discussing the pros and cons of the final results of his 35 years' search for a generalized relativity theory of a unified continuous field, he wrote that one could give good reasons showing that, and why, reality cannot be at all be represented by a continuous field.

Whitrow: Could you now tell us how this critical attitude of Einstein's which you have described has influenced your own work?

Popper: The Einsteinian revolution has influenced my own views deeply: I feel that I would never have arrived at them without him. In my view it is fundamental to science that it consists of theories which are tentative, or hypothetical, or conjectural. This means that any theory may be overthrown, however successful it may have been, and however well it may have been tested. There can be no theory more spectacularly successful than Newton's; but Einstein showed that even Newton's theory was only a conjecture. Thus, what Einstein's example may teach the philosopher is that science consists

25

of bold speculative guesses controlled by merciless criticism which includes experimental tests.

One point about Einstein which impressed me perhaps more than any other was this: Einstein was highly critical of his own theories, not only in the sense that he was trying to discover and point out their limitations, but also in the sense that he tried, with respect to every theory he proposed, to find under what conditions he would regard it as refuted by experiment. That is, he tried to derive from each theory predictions, testable by future experiments, which he regarded as crucial for his theory, so that if his predictions were refuted he would give up the proposed theory. Thus while he regarded all physical theories – not only Newton's but also his own – as tentative guesses which might always be superseded by better ones, and which therefore could never be verified, he made it clear that he found it most important to specify the conditions which would make him look at his own theories as refuted or as falsified. This attitude became the basis of my own thesis of the logical asymmetry between verification and falsification or refutation: of the thesis that theories cannot be verified, but that they can be falsified.

Following Einstein's example, I tried at once to find out the limitations of this doctrine, and I was able to show how it was always possible to evade a refutation. But I also showed that the possibility of such an evasion did not destroy the thesis of the logical asymmetry between verification and falsification. And I pointed out that the readiness to eschew such evasions and to accept falsification was one of the basic characteristics of the critical or scientific attitude.

Whitrow: Could you give us an illustration?

Popper: Yes. I may perhaps illustrate this point by an example from Einstein's own career. When D. C. Miller, who had always been an opponent of Einstein, announced that he had overwhelming experimental evidence against special relativity, Einstein at once declared that if these results should

26

be substantiated he would give up his theory. At the time some tests, regarded by Einstein as potential refutations, had yielded favourable results, and for this and other reasons many physicists were doubtful about Miller's alleged refutations. Moreover, Miller's results were regarded as quantitively implausible. They were, one might say, neither here nor there. Yet Einstein did not try to hedge. He made it quite clear that, if Miller's results were confirmed, he would give up special relativity and, with it, general relativity also.

This readiness to give up one's theory in accordance with the verdict of experiments is most characteristic of Einstein. It characterizes not only his critical or scientific attitude, but what may be described as his scientific realism. Although he knew that it was always possible to uphold one's theoretical constructions against unfavourable experimental evidence, he was not interested in doing so. He believed in some objectively existing reality which he tried 'to catch in a wildly speculative way', to use his own words: he was not content to find some equations fitting the observations, but he tried to grasp, to understand, this reality behind the phenomena. Yet he would have found this wild attempt uninteresting unless he could submit it to the discipline of rigorous experimental tests.

This attitude of Einstein is even today far from being generally accepted. Physicists and philosophers still speak of the verification of predictions, and even of the experimental verification of theories. But experiments have always to be interpreted in the light of theories, and theories can never be verified but remain always conjectures, wild attempts to grasp, or to understand, the hidden reality behind the phenomenal world.

Einstein's own views on the philosophy of science changed considerably during the course of his life. In his earlier writings there are many traces of positivist and conventionalist ideas. Especially noticeable is the influence of Ernst Mach, and also

27

that of the great mathematician Henri Poincaré, who was, indeed, one of the fathers of the special theory of relativity. Einstein said things which contributed much to the positivistic doctrines of 'operational definitions' and 'meaning analysis' – doctrines that were largely based on his own famous analysis of simultaneity. In his later years, however, Einstein turned away from positivism and he told me that he regretted having given encouragement to an attitude that he now regarded not only as mistaken but as dangerous for the future development of both physical science and its philosophy. He saw more and more clearly that the growth of knowledge consisted in the formulation of theories which were far removed from observational experience. I admit, of course, that we attempt to control the purely speculative elements of our theories by ingenious experiments. Nevertheless, all our experiments are guided by theory and they cannot be interpreted except by theory. It is our inventiveness, our imagination, our intellect, and especially the use of our critical faculties in discussing and comparing our theories that make it possible for our knowledge to grow.

28

Opposites Meet

I T WAS A HEROIC TIME in the full sense of the word. Old worlds, or their foundations, were being destroyed. New worlds would have to take form. At the moment though there was only the crumbling and sapping of the old, honored, and established precepts. It was of such upheavals that the old Greek myths had told. The word heroic thus was accepted by nearly all who lived through the period.

Human knowledge and its meaning were at stake. Quantum physics, in the light both of Heisenberg's quantum mechanics and Schrödinger's wave mechanics, was decreeing that man through science could never know the final, ultimate answers to the universe, if that answer meant absolute certainty about each action. Thus the base itself was shaken.

Since the days of the Greeks and most particularly since the time of Galileo and Newton the goal and method of science had been clear. The physical universe was determined in its course by all-embracing laws. If the laws could be discovered they would lead to the answer. This was the certainty and the method; in it science found unfailing guidance. Laplace even held that if there were a great enough "intelligence" the whole state of the universe in any one instant might be comprehended and its history forecast for all time to come.

The doctrine called causality—the certainty that cause produces effect—underlay the whole of science. It was the base of

53

knowledge and even the assumption of the child—if I count out five pennies I will know how many I have.

The abrupt discovery that this rule of the ordinary world did not hold in the atomic world, and that after counting out there would not be certainty, but rather uncertainty and indeterminacy, shook even the most resolute of minds. Heisenberg spoke of his "despair"; Bohr, of "alarm." To have made one of the foremost of advances and to have produced only a hazy, unalterable uncertainty at the very base on which the earth was built was almost more than the scientist could stand.

Bohr, who had led physics step by step into this dilemma, was most acutely aware that a way out must be found. Everything could not end in a blur. But Bohr could glimpse a solution of the impasse and, in fact, the way toward a greater understanding than ever before. At first, however, every attempt at this new approach met with defeat and frustration. In the cold, dark months of early 1927, he reached the point of exhaustion. He had to get away for a rest.

He and Mrs. Bohr went off to Norway to ski. Bohr was an excellent skier. As he zoomed down the snow-covered mountain slopes and felt the stimulation of the cold, crisp air, the concepts with which he had been struggling began to fall into place. The blockages disappeared.

By the time Bohr returned to Copenhagen two weeks later he had the basic answer to the paradoxes quantum physics had posed for the scientific world. Like the problem itself, Bohr's answer had implications that would extend far beyond the strict confines of science. It would in the end propose one solution for the bitter political and cultural divisions of the world. This, though, was for the future; in 1927 Bohr simply named his concept *complementarity*.

On the first morning of his return, Bohr raced down the steps of his apartment, across the graveled drive, and two at a time up the steps of the institute to find Heisenberg. He poured out his ideas. Heisenberg listened, disputed, and affirmed. It was the beginning of another of their intense work drives.

It was fortunate that Bohr felt rested, for during the next two months he and Heisenberg worked constantly. They would stop for a few walks in the park and Margrethe would drag them away for meals, but otherwise the work continued almost without interruption. Every attempt at interpretation was tested with every possible real or imagined experiment. It was the only sure-footed way through the immense subtleties and profundities with which they were dealing.

They also kept in the closest touch with Pauli. They needed his keenly critical view of what they were attempting. The detection of flaws or discrepancies was essential. Bohr prized nothing more than an objection that permitted him to clarify a point or to make a correction.

By spring the theory was ready, the theory that Dirac said "led to a drastic change in the physicist's view of the world, perhaps the biggest that has yet taken place"; that Oppenheimer called "the inauguration of a new phase in the evolution of human thinking"; and that John A. Wheeler, professor of physics at Princeton University, described as "the most revolutionary scientific concept of this century."

Essentially the theory to which Bohr and Heisenberg came was that there are two truths rather than one alone. Or, in other terms, that there can be two aspects, both true, and furthermore that the two together offer science and man a more complete view and understanding of the atomic world than either could offer separately, and thus in the end a clearer view of the visible world built out of the invisible substratum of the atom. Instead of division, Bohr showed, the parts, the divisions, the components can be combined into a harmony greater than that of the sections. Each separate, even contradictory, aspect complements the other.

Bohr began with one premise and two paradoxes. The premise was the well-documented one—the atomic world is not like the visible one. As we step through the looking glass, a new order appears, different in kind as well as degree.

This led immediately to the first paradox. Since Newton, science had rested on the assumption that a particle, be it electron or the planet Venus, can be observed in a certain place, and if its speed of movement is known, tracked certainly to the place it will reach in a specified time. The astronomer thus predicts a future eclipse of the moon, or a motorist driving to work at the permitted number of miles an hour can be certain if there are no upsets of arriving "on time." Almost every human move, if it is only reaching for the pencil lying on the table is consciously or unconsciously calculated in the same way, and Einstein in the theory of relativity emphasized that every observation or measurement rests on the coincidence of two independent events at the same space-time point. So profoundly is this general certainty built into the macroscopic universe that its existence is taken for granted, like breathing.

In the atomic world, however, an eclipse—if there were such an event in that world—could not be predicted to the moment; or if an electron started out on its rounds at a certain speed it would be impossible to say that it would arrive at a certain point at a certain time. An atomic hand reaching out—again if such a fantasy could be imagined—might only come close to the atomic pencil on the table. There might be a dismaying, disconcerting miss.

The reason, Bohr emphasized, is that the electron cannot be observed without a disturbance of it. The observer's beam of light, as Heisenberg had demonstrated, will buffet the electron around and alter its velocity. "Any observation regarding the behavior of the electron in the atom will be accompanied by a change in the state of the atom," said Bohr.

If on the other hand there is no disturbance, if no light beam or colliding electron touches the atom, there can be no measurement and no knowledge. And, Bohr pointed out, all knowledge concerning the internal properties of atoms is derived from experiments on their radiation and collision reactions. "This situation has far reaching consequences."

Equally staggering was the second great paradox with which Bohr had to deal—the wave-particle problem. By the strongest evidence the electron is a particle—a "concentration of energy and momentum at a single point of space and at any single instant of time." When an electron hit one of Rutherford's screens it gave off a scintillating flash of light. Something had struck. Or when it was hit by another electron the collision very much resembled a collision between two automobiles. Energy was transferred, and one or both were thrown off their course. Such a reaction unequivocally bespoke a particle, a thing of substance. But de Broglie and Schrödinger had shown that the electron also behaves as a wave. The two pictures were mutually exclusive and almost unthinkable and yet both indubitably existed. Physics was confronted with a fact and an impossibility.

The test experiments and equations devised by Bohr and Heisenberg showed that the velocity of an atom could be determined with precision. Or its position in space could be calculated with finality. It was only that both its velocity and position could not be found with accuracy. About its position at any particular point at any particular time there always had to be some degree of uncertainty.

It was the same with the wave-orbit problem. If an imagined, ideal microscope could be trained on the electron, and if it had sufficient resolving power to show the electron moving in its orbit, the same trouble would develop. In the act of observation at least one light quantum would pass through the microscope and would be deflected by the electron. The instant the electron was affected by the light quantum its momentum and velocity would change.

Thus there was no way of precisely observing the orbit of the electron around the nucleus. The waves that would normally be set off by the electron and that would be making their own orbit would be altered in their turn if the electron were jarred out of its movement by the light ray. It would never be possible to observe more than one point on the orbit of the electron.

But Bohr did not see helplessness in the quandaries that barred the way forever in the electronic world where the dart of light was as large as its quarry. "From the point of view taken here," he said, "just the renunciation forms the necessary condition for an unambiguous definition of the energy of the atom. We must consider the very renunciation as an essential advance in our understanding."

Both pictures could be used, Bohr saw. By employing both, by going from one to the other, a right idea could be obtained of the "strange kind of reality" in the atom. One, said Bohr using the key word, was complementary to the other.

The knowledge of the position of the particle thus might be considered complementary to the knowledge of its velocity, and the knowledge of the electron as a particle complementary to the knowledge of it as a wave. By knowing both with the greatest accuracy possible, a more complete description of experience, a new synthesis was possible.

"However contrasting such phenomena may at first sight appear," said Bohr, "it must be realized that they are complementary, in the sense that taken together they exhaust all information about the atomic object which can be expressed in common language without ambiguity."

By knowing both, Bohr maintained with revolutionary impact, a better description of experience—a new harmony—is attainable. Predictions could be made and checked by any scientist, Bohr further pointed out. They would not be predictions of exactly what would happen at any instant, but of the probabilities that certain things will happen. In the same sense no one can predict that some one individual will die at a certain moment, though there is no difficulty in predicting how many in 100,000 will succumb in a set period.

Thus Bohr resolved the appalling contradictions by synthesis. He concluded that complementarity is a requirement of the laws of nature and, far from being a hindrance, is an indispensable logical tool. Bohr was working out a subtle doctrine. It took words and attitudes out of their familiar contexts and re-used them in new strange ways. The staff at the institute, the first to be exposed to the new doctrine, found it almost impossibly hard to understand.

Bohr, happening upon a debate on the problem in the institute library, draped a leg across one end of the library table, and looking at the struggling students with an understanding grin, comforted them with a quotation from the Abbé of Galiana (1728–87.) "One cannot bow in front of somebody without showing one's back to somebody else."

The problem, Bohr noted, is as old as the Greeks. When Leucippus and Democritus proposed the atom as the indivisible, smallest unit of matter, the substance of which all else is made, they took it for granted that the coarseness of the sense organs would forever prevent the direct observation of individual atoms.

C. T. R. Wilson's construction of the cloud chamber, in which shooting electrons leave the marks of their passing in fine lines of water droplets, and the Geiger counter, clicking out the passage of electrons, had altered the early inaccessibility of the individual atom. Then and many times later Bohr was to say that the description of ordinary experiences "presupposes the unrestricted divisibility of the course of the phenomena in space and time," and the linking of all steps in an unbroken chain of cause and effect.

In the ordinary world too, Bohr emphasized, it is assumed that an ordinary measurement can be made without any effect on the object measured. The rough tearing of the cloth from the bolt does not impair the measurement of a yard. In the same way the physicist can turn his microscope on a piece of iron without in any way affecting it. In the eerie world of the atom it was different.

"The crucial point implies the impossibility of a sharp separation between the behavior of atomic objects and the interaction with the measuring instruments which serves to define the conditions under which the phenomena appear," said Bohr. Bohr argued that the new concept should not be difficult for humans. After all, he noted, "we are both onlookers and actors in the great drama of existence."

When the pressures of this demanding work became too heavy and there was no time for a weekend at Tisvilde, Bohr often went sailing. Like most Danes, living so close to the sea, he loved boating. In 1926 he and two friends, Niels Bjerrum, the chemist, and Holger Hendriksen, the xylographer, bought *The Chita*. During the next eight years they went on many trips, short or long.

With Bohr on board the discussions began swiftly. Even the sight of the moon on the water would prompt Bohr to ask why the reflection appeared as a streak rather than as a large patch, and in short order all of them were drawn into the debate. Bjerrum always said that Bohr had a remarkable ability to start his friends thinking and to make them feel much cleverer than they really were.

On one cruise *The Chita* stopped at Skagen. They all wanted to see a church that had been buried in the shifting sand. Bohr had always been skilled at throwing stones and skipping them on the water. He picked up a pebble and tried to throw it over the church. This proved easy. He then started trying to get the stone to fall outside the shutters over the peepholes in the first and second stories of the tower. This also was soon accomplished. Bohr then aimed at the small holes in the shutters and when he found that he could hit them, evolved the idea of throwing their walking sticks and making them stand in the sand outside the shutters. This worked. Bohr set the next goal—to knock the sticks down by throwing stones at them.

"Finally," said Bjerrum, "he got us to try to get the sticks to hang up there with the handles through the holes in the shutters and this really did succeed. The rest of us gave up the idea of getting them down again by throwing stones, but Bohr continued and was delighted when he finally succeeded."[1]

As Bohr rounded out his work on the theory that would give new directions to the whole of thought, he received an invitation to participate in an international congress of physics in Italy in September, 1927. It was to be held at the mystically lovely Lake Como in celebration of the 100th birthday of Volta.

Mussolini was at the peak of his power and Italy was determined to make the meeting an outstanding one. It was, in fact, planned as "one of the realizations of the Fascist regime," of which Mussolini and the newspapers were constantly talking. It was also to put Italy into the running in the suddenly important new area of theoretical physics.

Physics in Italy had been at something of a standstill in the preceding half century or more. Then, thanks to a physicist who was the only non-Fascist member of Mussolini's cabinet, a chair of theoretical physics was established at the University of Rome. A young Italian named Enrico Fermi was named to fill it.

To all who would listen, Fermi was expounding the quantum theory and the Bohr atom. As Laura Fermi relates in her fine book

[1] See the chapter by Niels Bjerrum in *Niels Bohr His Life and Work*.

Atoms in the Family Fermi was having trouble convincing the young group around him that matter and energy could be both particle and wave. They argued that such an outlandish contention was a dogma to be accepted only on faith. Fermi as the chief exponent of the "faith" thereupon was nicknamed "the pope."

Fermi, his friend Emilio Segré, who had just been persuaded to shift from engineering to physics, and Fermi's student Franco Rasetti, went to the Como conference. As the meeting opened, Segré whispered to his mentors: "Who is the man with the soft look and the indistinct pronunciation?"

"That is Bohr," they whispered back.

"Bohr, who is he?"

"Fantastic," Rasetti answered, "Haven't you heard of Bohr's atom?"

By the time the meeting ended, even an engineering student recently commandeered into physics knew about Bohr. Bohr had chosen as the title of his paper "The Quantum Postulate and the Recent Development of Atomic Theory." Behind this unrevealing title he made his first public presentation of his revolutionary theory of complementarity. It stirred the Congress as the Mistral sometimes roils the ordinarily calm waters of Como.

Bohr had begun cautiously and gracefully: "In a field like this where we are wandering on new paths and have to rely on our own judgment in order to escape from the pitfalls surrounding us on all sides, we have perhaps more occasion than ever to be remindful of the work of the old masters who have prepared the ground and furnished us with our tools."

His theory was presented in a language part physics, part philosophy. It was not, however, the language of any of the traditional philosophers, nor of their schools; it was not positivism, materialism, or idealism, though elements of each could be found within it. All of this only added to the growing unease. Bohr had not only destroyed the props of the whole structure of physics and science, but had done it in unfamiliar words and terminology. Anger and bafflement combined with the sense of shock. Schrödinger and Von Laue objected strenuously that Bohr's interpretation was neither convincing nor conclusive. They were entirely unwilling to sacrifice so much of the traditional base of physics. Others conceded that change was necessary, but disliked Bohr's ideas.

The widespread and vocal opposition was generally united only at one point—instead of turning to uncertainty, indeterminacy, and statistics, they insisted fervently that physics had to stand with "reality." The smallest part of the universe had to exist as objectively as a city or a stone, whether or not it was observed. They wanted a sharp, clear reality, not a haze or probability.

Einstein was not at the Como congress, and until the master was heard from, no one could be wholly certain in his own position. The question all over Europe was "What will Einstein say?" Would he demolish Bohr?

The fifth Physical Conference of the Solvay Institute was to follow almost immediately in October, 1927. Both Bohr and Einstein would be there, as well as nearly all others who were contributing to theoretical physics. Lawrence Bragg and Arthur Compton came from the United States. De Broglie, Born, Heisenberg, and Schrödinger all were to speak on the formulation of the quantum theory.

The subject was "Electrons and Photons." To leave no doubt that it was directed to the main question, the theme embroiling all of physics, discussion was centered around the renunciation of certainty implied in the new methods. There lay the rub. The stage was also set for a discussion of the possibility that wave mechanics—interpreting the structure of the atom primarily in terms of waves—might offer a way for keeping the cherished solidity of the old and combining it with some of the new. Such a course would have entailed a far less radical departure than Bohr's complementarity.

Against this background, Bohr was invited to give the conference a report on the epistemological problems confronting quantum physics. By asking him to speak on the science of knowledge and the grounds for it, the conference gave him full opportunity to present complementarity. There was no avoidance; the issue had to be directly faced.

Excitement mounted as Einstein rose to speak. He did not

keep them long in suspense. He did not like uncertainty. He did not like the abandonment of "reality." He did not think complementarity was an acceptable solution, or a necessary one. "The weakness of the theory lies in the fact that on the one hand, no closer connection with the wave concept is obtainable," he said, "and on the other hand that it leaves to chance the time and the direction of the elementary processes."

A dozen physicists were shouting in a dozen languages for the floor. Individual arguments were breaking out in all parts of the room. Lorentz, who was presiding, pounded to restore order. He fought to keep the discussion within the bounds of amity and order. But so great was the noise and the commotion that Ehrenfest slipped up to the blackboard, erased some of the figures that filled it, and wrote: "The Lord did there confound the language of all the earth."

As the embattled physicists suddenly recognized the reference to the confusion of languages that beset the building of the tower of Babel, a roar of laughter went up. The first round had ended.

What was euphemistically called the "exchange of views" was continued in smaller groups during the evenings. And here Bohr and Einstein met directly, face to face. Ehrenfest, who for many years had been a friend of both Bohr and Einstein, in effect served as mediator. His services were not needed in the sense of preventing personal combat. This was a battle fought with soft words and courtesies. Each held his opponent in a respect verging on awe. But the differences were as fundamental and as sharply defended as though the weapons had been guns.

Bohr had prepared carefully. He never left his words to last-minute inspiration, but compiled his material and tested it out against all possible loopholes in discussions with associates and students at the institute.

Bohr hoped that his proof would bring Einstein to his side. Complementarity, he argued, only carried on ideas that Einstein himself had ingeniously raised. Was it not Einstein who in 1905 had shown that the photon—a ray of light—is a corpuscle as well as a wave? Had he not explored novel procedures outside the classical framework of physics to do so?

And had not Einstein himself in 1917 formulated the rules indicating that the atom may spontaneously emit radiation at a rate corresponding to a certain prior probability? Only the probability of that disintegration could be calculated, and not the exact moment at which the radioactive material would give off another bit of radiation, in its long or short decay.

And did he not emphasize the dilemma still further by showing that the radiation will flash only in a certain direction, though in the wave picture of radiation "there can be no question of a preference for a single direction in an emission process?"

Einstein had even concluded his paper on radiation by saying: "These features of the elementary process would seem to make the development of a proper quantum treatment of radiation almost unavoidable."

A man who had laid this foundation should not quibble about the similarly drastic changes in foundation proposed in the theory of complementarity, Bohr argued.

Einstein also was prepared. He was ready to show by "ideal experiment"—one of his imagined but correctly set up experiments—that uncertainty would not be necessary if the right experiments were made, and if the interaction between atomic objects and the measuring instruments was more explicity taken into account.

Einstein went to the blackboard. He drew one line with a small slit in it, and just beyond it another line representing a photographic plate. If a single electron or photon went through the slit, it would as it emerged fan out as a typical wave, in the wave's concentric lines. Einstein conceded that it would not be possible to predict with certainty at what point the electron would arrive at the photographic plate, say at Point A in the upper part of the plate, or at Point B, in the lower part. It could only be calculated that the electron *probably* would be found within a given region of the plate. But, Einstein argued, if the electron is recorded at Point A, it could never be recorded at Point B. Thus it should be possible, with control of the momentum and energy transfer, to determine precisely where and when the electron would strike.

Bohr objected immediately that the experiment Einstein had sketched was not comparable to the application of statistics in dealing with complicated systems. They quickly came down to the question—does the quantum-mechanical description exhaust the possibilities of accounting for observable phenomena? Einstein's answer was a decided "No."

Bohr challenged Einstein to examine "the simple case" of a particle penetrating through the hole if there were a shutter to open and close the hole. Bohr went to the blackboard. He added several lines to indicate a shutter. Could the shutter be used to control the momentum and energy transfer involved in the location of a particle in space and time? Could the collision of the light particle be controlled enough not to interfere with the course of the electron under study?

Bohr's answer was that the moment the light was weakened enough not to affect the electron, the electron could not be clearly found. It would become impossible to fix both its place and velocity.

Einstein was not in the least satisfied. "I am firmly convinced," he said, "that the essentially statistical character of quantum theory is solely to be ascribed to the fact that the theory operates with an incomplete description of physical systems." He was insisting that Bohr had still not gone to the bottom, and that he was taking an incomplete answer for the final one.

At their next session Einstein was ready with still another "imaginary experiment." Between the diaphragm with the slit and a photographic plate, Einstein inserted another diaphragm with two parallel slits.

Einstein suggested that proper control would make it possible to decide through which of the two slits the electron had passed before striking the photographic plate and registering its arrival with a black spot.

The single electron could go through only one of the two slits—after all it could not split in two. And the electron would strike the photographic plate at a predictable time and spot. Einstein argued that it would be possible to decide through which of the two slots the electron had passed.

This was not an easy one, but Bohr rose to it. After deep thought he demonstrated that in controlling the second diaphragm there would again be an inevitable interference with the electron and again indeterminacy would come into play.

"We are presented," said Bohr, "with a choice of either [and he emphasized the either] tracing the path of a particle or [again great emphasis] observing interference effects. We are faced with the impossibility of drawing any sharp separation between an independent behaviour of atomic objects and their interaction with the measuring instruments."

Bohr thought that he had proved his point. But Einstein was no more convinced than when the discussions began. If the arrangements suggested were not sufficient to obtain accuracy, he was certain that others should be possible. It was only lack of knowledge that blocked the way, not an inevitable barring of the way in. Einstein was not willing, because knowledge was lacking, to fall back upon uncertainty. He would not consent to be satisfied with saying "There's a chance that it will be this way," instead of "This is the way it will be."

"In spite of all divergences of approach and opinion," said Bohr, "a most harmonious spirit animated the discussions."

Einstein lightly jibed at Bohr and his support of chance and probability: "Do you really believe God resorts to dice-playing?" Bohr came back in the same spirit: "Don't you think caution is needed in ascribing attributes to Providence in ordinary language?"

Ehrenfest, though he was Einstein's collaborator, was shaken by Bohr's arguments. He teased Einstein, hinting that Einstein's attitude toward complementarity and the new development of the quantum theory was similar to that of opponents of relativity. Nevertheless Ehrenfest did not want to imply that he had switched to Bohr's side. With his next breath, he added that he could never feel any peace of mind about complementarity unless Einstein were convinced.

Despite Bohr's conviction that he had demolished Einstein, Einstein's insistence that there should be a way made him question. Again it was the kind of challenge Bohr loved. Could experiments possibly be devised that would enable the scientist to discover the particle at a particular spot and time?

He and the Copenhagen delegation worked most of the night trying to imagine other instruments and experiments that would control the momentum of the particles enough to make them fully measurable. But the more they tried that night—and in later years—the more impossible the task looked.

Einstein also persisted. In one of the general meetings Einstein again raised his general objections. The quantum attempt to solve the riddle of the double nature of all corpuscles had not found a final solution in the statistical quantum theory.

Einstein considered the impasse, as Poincaré had once said, "a measure of the depth of our ignorance," and he repeated his even greater dissatisfaction with another phase of the quantum theory, its failure to get down to reality. It did not give a complete description of the real, individual situation, irrespective of any act of observation or substantiation. Look at the radioactive atom, Einstein urged. At a certain time it emits a particle. The individual atom therefore has a definite disintegration time.

"One is driven to the conviction that a complete description of a single system should after all, be possible," Einstein declared. "But for such a complete description there is no room in the conceptual world of statistical quantum theory."

Bohr answered *yes*, but the consideration stands or falls with the assertion that there actually is such a thing as a definite time of disintegration of the individual atom. "The assertion of the existence of a definite time-instant makes sense only if I can in principle determine the time-instant empirically," he said. To determine, Bohr went on, would involve a definite disturbance of the system.

Einstein did not like this any more than when he heard it for the first time. He accused the quantum advocates of being "egg walkers," willing to go to almost any length to avoid the "physically real." Einstein was particularly irked by Bohr's general contention that a search for a complete description would be aimless, for the reason that the laws of nature can be completely and suitably formulated within the framework of the quantum description.

But, said Einstein, pounding as hard as his gentle disposition permitted, "for me the expectation of the adequate formulation of the universal laws involves the use of *all* conceptual elements."

Einstein hit hard again: "It is not surprising that, using an incomplete description (in the main), only statistical statements can be obtained out of such a description." Incomplete data, he was charging, produce incomplete results.

"To me it seems a mistake to permit theoretical description to be directly dependent upon acts of empirical assertion, as it seems to me to be intended in Bohr's principle of complementarity."

Solvay, 1927, one of the most influential conferences in the history of science, then ended. Physicists returned to their laboratories all over the world to continue the argument and to work on the problem.

The heroic period in physics, the unmatchable period, also was coming to an end.

Scientists Are People

Raymond J. Seeger

A major problem of modern education is the general *drop-out* from school: of the students in fifth grade a few years ago about 72% will graduate from high school and 40% will enter college, but only 20% will graduate from it. A minor problem is the special *stay-out* from certain disciplines, particularly physics. During the past five years the number of college students receiving a bachelor's degree in physics has decreased 10%. Over the past 15 years even the high school enrollment in physics has decreased 1%, whereas in science, generally, it has increased 9% and in mathematics 11%. It is sometimes argued that the mathematical requirements for modern physics are a deterrent for students contemplating this subject. If so, then the very increase in students taking mathematics in secondary schools should alleviate the problem instead of aggravating it. Evidently the primary need now is better coordination between mathematics courses and physics courses in high school, with more emphasis upon applications of mathematical reasoning to understanding natural phenomena.

Although we are not particularly concerned in this article with the factors that contribute to the physics stay-out, in passing, I should like to call attention to a few. In the first place, there is the stereotype image of the scientist. In an opinion poll[1] about one-third of the high school students blamed the difficulties encountered in science on their own poor academic background coupled with the apparent inadequacy of their intellectual capacity (the supposed need to be a "genius"). They complained, furthermore, of scientists being a group of "odd" people, who jealously guard their own interests and fail to sacrifice for other people, who have no time to enjoy life and can't even have a normal family. Ten percent of the students claimed that scientists are inherently dishonest and evil. These conclusions, unfortunately, are abetted by the narrow opinions of many professional counsellors, whose guidance too often reflects merely their own lack of adequate scientific experience. Of equal importance are the careless actions of many physicists who frequently fail to communicate the spiritual depth of their own research interests. What is of most concern to me, however, is the lack of any evident improvement in such attitudes from the first year of high school through the senior year.[2] In view of the relatively few students taking physics during this educational phase, physics teachers cannot be entirely to blame. Perhaps teachers of biology and of chemistry have an even greater responsibility—not to mention the influence of teachers in other disciplines inasmuch as they have these students for most of their high school career.

A second stay-out factor is the practical one of scheduling. Most high school biology today is obviously more or less a compromise between traditional natural history and modern molecular biology. It is usually based upon modern chemistry, which some of the students get in their junior year. (The chemistry, in turn, is based upon modern physics, which the student gets only in his senior year—if at all.) This sequence is obviously the reverse of any educationally sound approach to modern biology, which is founded largely upon biochemistry and biophysics.

The perennial need for motivation is highly significant; its correction, therefore, will afford some degree of a practicable solution to the stay-out problem. Let us briefly analyze this factor from the point of view of science as experience combined with reason and imagination. Of primary importance is developing curiosity about phenomena themselves. Unfortunately, modern urbanization with its increasing loss of blue skies and green fields restricts opportunities for people even to notice particularly interesting things. At the same time, the complex gadgetry in "black" boxes does not always invite an inquiring mind in school. Sometimes, however, motivation is associated more with the eternal quest for comprehensive understanding, for an over-all view. In this instance there is unquestionably a potential for stimulating student interest, inasmuch as theoretical concepts are nowadays emphasized more frequently than empirical approaches. An over-emphasis, however, upon the status quo is also a danger signal, warning that it is not only important to know where we are, but also where we expect to be. To get a sense of direction, indeed, we need at least two points; the one where are we and the other, where we were: the line joining them indicates a probable future outlook. Students must be taught not merely to understand the science of today, but even more to be prepared for its inevitable change 25 years hence. The science of tomorrow, we can be certain, will be as different from today's science as the latter is from the science of yesterday. A third factor in motivation is the fellowship of imaginative persons (living and dead). Some years ago I came across a book with the interesting title, *Children Are People*.[3] So, too, I mused, students are people, teachers are people, scientists are people! What I propose, therefore, is an emphasis upon people-to-people communication, one to another. Let us not try to classify all students, all teachers, all scientists, but rather regard each one as an individual person. I am convinced that the most important contribution a high school teacher can make to education is the personal motivation of students. You may recall the prophet Jeremiah brooding about the ruins of Jerusalem. What would happen now, he wondered, to the promised land that had just been destroyed by the conqueror, to the promise made to the people recently taken captive to Babylon? It occurred to Jeremiah that, although the covenant between God and the people of Israel could not now be in effect owing to the very dissolution of the nation, nevertheless, through individuals, the existent remnant, a new covenant could be made—this time between God and each person. In this spirit, therefore, I propose a teenager-to-teenager communication, a bridge between young people of today with those young people of yesterday who became scientists, who themselves had seemingly formidable problems as growing persons. We shall find that their predicaments were not too different from those encountered by teenagers today. Accordingly, I have selected, somewhat at random, 15 well-known scientists, and have grouped them roughly into the advantaged, the undecided, and the frustrated. Let us look at each of these scientists when he was a teenager.

The first of the advantaged I shall call genius "X". He was born in 1623 in Clermont near the volcanic Auvergne. His father was a magistrate who went to live in Paris in 1631, but had to go into hiding (1638) because of certain demonstrations against the government. Upon being pardoned later that year he became a tax collector at Rouen. The mother, who had some interest in mathematics, died when the child was three. The family were Jansenists. As a baby the boy was sickly; his abdomen was puffed and hardened. (After 18 years of age he is said to have spent no day without pain.) His early instruction was given entirely at home, where his father taught him grammar, ecclesiastical history, and some science. (Latin and mathematics were deferred until a later time.) The boy benefited much by self study. Despite his father's restriction on geometry, at 12 he was found ferreting out theorems with his own charcoal diagrams. At 16 he wrote a one-page treatise on conic sections, in which he utilized the then new mathematics (projective geometry) of Gerard Desargues (not fully appreciated until 200 years later). At 19 he had to help his father with calculations for tax assessments. Anxious to escape this drudgery, he designed the first calculating machine (10 of the 50 constructed are extant). His scientific achievements were outstanding. At 24 he made major contributions to the physics of fluids and at 31 laid down the foundations of probability. Meanwhile, however, he had dedicated his life wholly to religion. His writing of the "Lettres provinciales" at 33 has earned him recognition as the father of modern French prose. Having to wait on occasions with others for individual carriages, at 35 he conceived the omnibus (for all), a suggestion which was taken up and put into practice—at five sous a ride. He died at 39. This scientist "X" was truly a genius, this religious man of science—Blaise Pascal.[4]

We shall consider precocious "Y" as our second advantaged person. He was born in 1773 at Milverton (near Taunton, Somerset). The boy was the first of ten children. His father was a mercer (cloth merchant), a banker, and a Quaker. He learned to read at the age of two (he actually attended a village school then). By four he had read the whole Bible twice, together with some of Isaac Watts' hymns. At five he could recite Oliver Goldsmith's "The Deserted Village." At six he began his study of Latin. He was then sent to a boarding grammar school at 6 plus, but became homesick. Returning home he read about science in the library there. At 8 plus he again went away to boarding school, where he heard his first lectures on natural philosophy and where he enjoyed making telescopes and binding books. By 13 he would read Latin and Greek, French and Italian, as well as Hebrew. At 14 he was hired as a tutor—companion for a Barclay grandson. In this capacity he read English, French, and Italian classics. (He translated some English classics into Greek.) He also read Isaac Newton's "Principia" and "Opticks." At 16, stirred by the Negro trade, as a Quaker, he abstained from sugar. At this stage modern guidance counsellors would undoubtedly have recommended that he embark upon a literary career. During that very year, however, he became ill and was treated by an uncle (Dr. Brocklesby), who persuaded him to study medicine. The boy attended lectures on chemistry and performed some simple experiments on his own, although he was never particularly interested in experiments per se. At 19 he began his higher education in the Hunterian School of Anatomy in London. At 20 he entered St. Bartholomew's Hospital and gave a paper on "Observations on Vision," which was published by the Royal Society the following year. (He was elected a fellow of the Royal Society.) He then continued his medical education at the University of Edinburgh. Because of his developing interest in music, dancing, and the theater, he separated himself from the Quakers. At 22 he went to the University of Göttingen (founded in 1733 by George II). Here his curiosity was aroused about sound, particularly its production and propagation —the beginning of his interest in physics. At 24, a converted Anglican, he entered Emmanual College at Cambridge to receive its prestigious degree (at 25) preparatory to becoming a practising physician. As for his scientific accomplishments, at 27 he became Professor of Physics at the Royal Institution (1800-1802). His lectures there on Natural Philosophy were published in 1807; in them he argued convincingly for the wave theory of light, based upon a thought experiment involving the phenomenon of interference. Meanwhile, the Egyptian Rosetta stone, discovered by France in 1799 and obtained by England in 1811, presented an unusual problem of linguistic interpretation, inasmuch as it was written in some undeciphered languages: hieroglyphic, Egyptian, and Encorial inscriptions. This scholar made some translations of them; six turned out to be correct, four partially correct, and four wrong. He died at 56. A medallion commemorates him in the Westminster Abbey. This scientific man of medicine "Y" was truly precocious—Thomas Young.[5]

Our last instance of an advantaged person is a gentleman and scholar "Z". He was born in 1831 at Edinburgh. As Laird of an estate inherited from a brother, the father was a sportsman with practical interests, but at the same time quite unworldly (he was an elder in the Scottish Church). His mother, whose father had been prominent, was a gentlewoman, a pious Anglican, who composed for the organ, but who unfortunately died when the boy was only nine. Being an only child he enjoyed a close relationship with his father. In the pregrammar school period the boy was wont to ask, "What's the go o' that?" At eight he had some knowledge of the Scriptures and Milton; he could repeat the lengthy 119th Psalm. He was accurate in drawing. While at home he was accustomed to wandering. At ten he went to a Scottish day school in Edinburgh, where he lived with an aunt. He had to attend both St. Andrew's Presbyterian Church and St. John's Anglican Church. Shy and dull at school, he consequently had few friends, particularly in view of his lack of interest in sports. At 13 he made some models of the five regular solids and became interested in geometry. Although he was only 11th in the list of scholars the next year, he obtained first prize in English, as well as a prize for English verse and a mathematics medal. At 15 he wrote a paper on "Oval Curves," which was read at the Royal Society of Edinburgh. He did not receive a medal that year, but did become interested in magnetism and electricity. At 16 he received first prizes in English and in Mathematics (and almost in Latin). His higher education began with his entrance to the University of Edinburgh at 17. Here he studied logic, mathematics, and natural philosophy. At 18 he wrote a paper on rolling curves, and at 19, one on elastic solids. During that year he entered Peterhouse College at Cambridge (he should have probably gone there a year earlier). He later transferred to Trinity College, where the Master was William Whewell. He did not, however, receive a scholarship until 21. He was noted for his verse, usually humorous, sometimes ironical; nevertheless, he wrote a "Student's Evening Hymn." At 23 he was awarded the distinction of second Wrangler and tied with the first Wrangler, Edward John Routh, for the coveted Smith prize. By this time he had developed his life-long interest in electricity and magnetism, as well as an ingenious color top. At 24 he was elected a fellow of Trinity College. Although always receptive as a student, he managed to put his own stamp on all his work. Later he always acknowledged appreciatively his education. He became Professor of Physics at Marischal College (Aberdeen), then at King's College (London, 1860–65), where he had contacts with Michael Faraday. Later he was called to the University of Cambridge, where he was responsible for building the world-famous Cavendish Laboratory. He received the Adams prize for his essay on Saturn's Rings, was one of the founders of the kinetic theory of gases, and wrote the classical "Treatise on Electricity and Magnetism" (1873), in which he summarized previous experimental findings in mathematical terms and formulated the electromagnetic theory of light—confirmed experimentally in 1888. This professor of experimental physics "Z" was truly gentleman and a scholar—James Clerk Maxwell.[6]

Let us consider a few cases in the undecided class. We shall begin with ministerial student "A". He was born in 1571 at Weil der Stadt. A Lutheran grandfather had been a bookbinder and mayor, but was apparently never particularly interested in this child. His father at first assisted the grandfather, but then devoted most of his life to soldiering; he was rude and rough to his wife and actually abandoned his family in 1583. His mother in later life was jailed for 14 months and then tried as a witch; she died obscurely. A brother (epileptic) was a good-for-nothing and ran away; another, however, was a respectable craftsman (a pewterer); a sister married a clergyman. The child himself almost died with smallpox while his parents were away engaged in war in the Netherlands. His early education was in a German school in Wurttemberg. At seven he attended a Latin school, where continual moving and hard agricultural labor necessitated his spending five years to complete the normal three-year course. He did, however, show evidence of a keen intellect, primarily in the state examinations. At 13, accordingly, he was enrolled in a convent school at Adelberg. At 15, he was admitted to a higher seminary located in an old Cistercian Monastery. He was a conscientious student, but quite introspective. He exhibited keen spiritual anxiety and became increasingly disturbed about religious controversies. He was always busy, but unable to stick long at any one thing. At 17 he passed the entrance examination for the University of Tübingen on the Neckar, but returned for another year to the seminary at Maulbronn. At 18 he entered the University, specifically the Stift Seminary, where his life was wholly regulated. In philosophy there he preferred Plato to Aristotle, and became interested in the mystic Nicholas of Cusa. A teacher, Michael Maestlin, stimulated his interest in mathematics and astronomy, particularly in Euclid and Apollonius, Archimedes, Ptolemy and Copernicus. He became increasingly discouraged in theology, owing to constant squabbling; indeed, he developed a bitter distaste for all such controversy. At 23, the faculty unexpectedly recommended him as a mathematics teacher to the Graz Seminary (protestant). His parents would have preferred the prestige of a priest, but deferred to the advice of the faculty. At this time, he abandoned his native land forever. Later he became assistant to the celebrated astronomer Tycho Brahe at Prague, and eventually himself became Imperial Mathematician. He completed the Rudolphine astronomical tables and formulated the three laws of planetary motion. In 1604 he detected a new star (nova); he later studied optics with particular interest in the design of telescopes. On the side, however, he maintained a practical interest in astrology. This undecided ministerial student "A" became the diligent mathematical astronomer, Johannes Kepler.[7]

Let us next consider a premedical drop-out "B". He was born at Pisa in 1564 (a memorable year for births and deaths of celebrities). His father was of an impoverished lower nobility and had to resort to trade (cloth) for his living. The father, however, was interested in language, mathematics, and music, particularly the lute. The boy went initially to school at Pisa and with his father's help studied some classics. He made toy machines, which did not always work. He then attended the monastery school at Vallombrosa, where he learned to enjoy the classics and

Italian Poetry, particularly Dante, Petrarch, and Ariosto. He learned to play the lute and organ. He liked to draw and paint; he admitted later he would have chosen art for his life's work had he been free to do so. At 17 plus, as a premedical student, he entered the University of Pisa. Overhearing a mathematics tutor, at 18 he found himself suddenly fascinated by mathematics. With the consent of his father, he decided not to study medicine. Being attentive to the lectures was not enough to win the professors' good will and a much needed scholarship, inasmuch as he was also quite argumentative. At 20, therefore, owing to financial straits, he had to return to his family at Florence; he never graduated. At 21 he wrote a paper on the center of gravity, and at 22, one on "The Little Balance." At 25 he was appointed Professor at the University of Pisa, the very institution from which he had dropped out. He later became Professor at the University of Padua and finally Chief Mathematician for the Grand Duke of Tuscany. He is largely responsible for the so-called scientific method as we know it today, for the application of mathematics to observations, for the interaction of experiment with theory. He was unqualifiedly the founder of dynamics and acoustics, as well as the first to design a thermoscope, a telescope, and a compound microscope. He was always a sharp, aggressive opponent of all authoritarianism, both in philosophy and theology; he was an enthusiastic proponent of the Copernican system. This undecided premedical drop-out "B" became the founder of modern physics, Galileo Galilei.[8]

We now come to a farm boy "C", who was born on Christmas at Woolsthorpe, seven miles south of Grantham, the very year that Galileo died. His father, who had a small manor (a poor estate), died three months before the boy was born. Within two years the mother married a clergyman, by whom she had three children. Meanwhile the boy had to live with his grandmother and an uncle. As a premature child, he was quite frail; in his early years he had to wear a bolster about his neck to support his head. He attended day schools at Skillington and Stoke. At 12 he entered King's School at Grantham, where he was placed in the lowest form, indeed, next to the last in his class. It is said that the boy who ranked above him one day kicked him in the stomach. He fought the boy and determined to beat him also in studies. Ultimately he became first in school. Nevertheless, he showed no unusual ability. At 16 he had to return home to help farm and manage the estate (his stepfather had died in 1656), but he was quite negligent of the animals—a total failure at farming (he studied mathematics while supposedly at work). At 18, accordingly, he returned to King's School to prepare for Cambridge, where he entered Trinity College the next year; a shy and diffident country lad; he found it populous and busy. His first years were spent without distinction, undoubtedly owing to the keen competition of the better prepared public school boys. As a sizar, moreover, he had to work for his room and board by performing menial services. He was not popular; on the one hand, he was shunned by fellow students because of his lack of interest in physical exercise and sports; and, on the other, he was not attractive to older men, inasmuch as he was not sociable, talkative, or witty. At 21 one of the teachers, Isaac Barrow, gave some lectures on natural philosophy, including optics, which greatly interested the young man. Together with 44 other students he received a scholarship, and at 22 obtained the bachelor's degree—without any distinction being noted. He returned to Woolsthorpe, where he sat out the great (black) bubonic plague until the University re-opened. It was during this time that he meditated on calculus (fluxions), the nature of white light, and universal gravitation. His outstanding scientific accomplishment was the publication of the "Philosophiae Naturalis Principia Mathematica" in 1687, which included the fundamental laws of motion, and which has been acclaimed as "the greatest intellectual feat in the history of science." In 1703 he wrote a book on "Opticks." Meanwhile, at 26 he had been appointed Lucasian Professor of Mathematics at Trinity. Later, he became warden of the mint, then its master, a member of Parliament in 1701, President of the Royal Society in 1703; he was knighted two years later. He died at the age of 84 and was buried in Westminster Abbey. This undecided farm boy "C" became the foremost theoretical physicist (and mathematician), Isaac Newton.[9]

Let us now consider the musician deserter "D", born at Hanover in 1738. His father was an oboeist in the Guards. His mother had ten children, of which he was the fourth. She was opposed to learning and made it impossible for her daughters to learn French or dancing. She herself could not write. The boy attended a garrison school until 14, where he learned some French and later, on the outside, Latin and arithmetic. His father taught him to play the oboe, the violin, and the organ. At 17 he, too, became an oboeist in the Guards and toured England for a year. He participated in the campaign of 1759, but the next year, at 19, he decided on "removal" (desertion), ostensibly because of poor health; penniless, he returned to England. (On his first official visit to the King of England in 1789, he was presented with a formal pardon.) At 22, he took charge of the music for the militia of Durham. Essentially a free lance, he found himself involved with pupils, concerts, and compositions. At 28 he became orga-

nist at the octagon chapel in fashionable Bath, where he
was engaged in both composition and teaching. By himself
he studied Italian, Greek, and mathematics. He became
interested in the applicability of mathematics to optics in
general, and then to astronomy in particular. At 36 he
made a Gregorian telescope. A year later he used a
Newtonian telescope to survey the whole sky and to locate
planets and stars above the 4th magnitude. At 41 he
joined the new Philosophical Society at Bath, and at 44
decided to devote full time to astronomy. Scientifically, his
most notable discovery was the planet Uranus in 1781,
the first since "shepherds watched their flocks by night."
Subsequently, he was made a fellow of the Royal Society
and received the Copley medal. At the same time he
continued to identify nebulae and star clusters; he proposed a structure of the universe. He was knighted in 1816
and died in 1822. A memorial stone has been set in
Westminister Abbey's floor near his son's tomb. This undecided musician deserter "D" became the astronomical ex-
plorer, Frederick William Herschel.[10]

Next we shall consider an adventurer "E", born 1753 at
North Woburn, Massachusetts. His father was a fifth-
generation American farmer. Unfortunately he died when
the child was only two; the mother remarried. The boy
went to school until 13. Nothing of genius was evident;
indeed, he was said to be "indolent, flighty, unpromising."
With the aid of an older boy, however, he did manage
some self-education. At 14 he became an apprentice to a
dry goods merchant in Salem, where he exhibited good
draftsmanship. At 16 he transferred to a dry goods merchant in Boston, where he once walked to Harvard to hear
Professor John Winthrop lecture on natural philosophy.
He tried to make a perpetual motion machine. He is said
to have participated in the Boston "massacre" (1770).
The fireworks he prepared for a Stamp Act celebration
blew up and badly burned him. His employer, concerned
about such a risk in his store, fired the boy. He became an
itinerant teacher. At 19 he went to Rumford (later
Concord), New Hampshire, where he taught for a clergyman. An aggressive daughter, a 32-year old widow with
the largest fortune in town, arranged a marriage with the
adventurer. They had only one child. He himself used his
position to become commissioned by the Governor at
Portsmouth as a major in the Second Provincial Regiment. His aristocratic airs, however, earned him the epithet "dandy and upstart;" he was unpopular and at 22
was regarded as unfriendly to the popular cause of liberty.
(He was actually a spy, and from time to time sent
information to General Thomas Gage in Boston.) He was
"tried" but found "not guilty." Nevertheless, there was

considerable agitation to have him tarred and feathered. He escaped—never to return. At 23 when the British evacuated Boston he went with a group to England. Owing to some successful experiments with gunpowder, at 27 he became a fellow of the Royal Society. The next year he was made a lieutenant colonel in the British Army and returned with its New York regiment to America. A year later, as a colonel he returned to England. At 31 he went to serve the Bavarian Elector at Munich. Meanwhile, he had been knighted by George III. At 32 he returned again to Munich and was made there a Count of the Holy Roman Empire of the German nation. At 52 he married Antoine Laurent Lavoisier's widow. He died in France at 61. His scientific achievements were not of a major character, although he did make a significant contribution to the understanding of heat. He was primarily important for his organizing activities; for example, he was one of the founders of the Royal Institution in London (corresponding to the later Smithsonian Institution in Washington, established by an Englishman) in 1799. This undecided adventurer "E" became the well-known physicist, Sir Benjamin Thompson, Count Rumford.[11]

We come now to a country gentleman "F", born at Shrewsbury in 1809. His father, a successful country physician, married a daughter of the potter Josiah Wedgewood. They had six children, of whom he was the fourth. The mother died when he was eight, so that he was brought up mainly by his older sister. In the same year he was sent to day school (Unitarian), where he was slower at learning even than his younger sister. He did, however, show interest in collecting plants and eggs, pebbles and shells. His father's judgment at that time was the following: "You care for nothing but shooting dogs, and will be a disgrace to yourself and your family." At 9 he went to school under Samuel Butler, grandfather of the author of "Erewhon." Later he himself reflected that "the school as a means of education to me was simply a blank." At 16 he went to the University of Edinburgh to study medicine. He found the lectures there incredibly dull, with the exception of those on chemistry. He was actually rebuked by another student for investigating biological phenomena in what the student regarded as his own pre-empted field Aware of his father's property, the young man made no great effort to prepare himself for a career in medicine. His father, therefore, proposed holy orders. And so at 18 he entered Christ College, Cambridge. He regarded his years there "wasted" as far as academic studies were concerned, "as completely as at Edinburgh and as at school." He continued shooting and hunting—and collected beetles. The Reverend John Steven Henslow, however, stimulated his interest in botany. At 22 he graduated 10th among those not seeking honors. He did, however, make a geological excursion into Wales with Professor Adam Sedgwick. He then sought a job as naturalist aboard the ship Beagle. His father, however, questioned such an experience for a potential clergyman, owing to the disrepute and discomfort associated with it. An uncle, however, argued that natural history may not be unsuitable for a clergyman. The young man, meanwhile, had been rejected by the ship's captain owing to his physiognomy (a poor nose shape did not seem suitable for the hardships that would be encountered), but was later accepted. He determined to make collections on the journey. At 27 he returned to England eager to pursue an entirely different career, namely, that of a man of science. He was elected a fellow of the Royal Society at 29 plus. Having poor health he elected to live in a country house (Downs) near Seven Oaks in Kent with a Wedgewood daughter, whom he married in 1839). His publications, "Journal of Researches" (1839), "On the Origin of The Species" (1859), and "The Descent of Man" (1871), all have had wide-reaching significance. He received the Copley Medal in 1864. Although buried in Westminister Abbey in 1862, he was never honored otherwise by the Government. This undecided country gentleman "F" became the revolutionary biologist, Charles Darwin.[12]

We shall conclude this group of the undecided with runaway "G", born at Boston in 1706. His father, an emigrant from Banbury, England, was a chandler. His mother, a second wife, had ten children. The boy learned to read early and attended what is now known as the Boston Latin School; he was head of his class. Noting the ultimate cost of college education, even in those days, the father withdrew the boy at nine and sent him to a day school where he learned to write, but failed in arithmetic. At ten he was withdrawn from this school, too, to assist his father and to learn to handle tools. At 12 he was apprenticed to a printer (his brother) to learn this trade. He was fond of reading; he borrowed books and even bought some of them with money saved on food. At 16 under a pseudonym he wrote for a newspaper. At 17 he sold some of his books and set out to seek his fortune elsewhere. He walked to New York and from there to Philadelphia, where he arrived with one Dutch dollar and one copper shilling. Whereas at Boston his studies had been his major concern, in Philadelphia his interest turned to friends (he married there at 24.) At 18 he made a trip to London, where he became stranded and had to practice the trade of printing to obtain funds. Returning at 20, he made some quantitative observations of oceanic properties on this trip. It was not, however, until the age of 40 that he had sufficient leisure to pursue freely his curiosity about natural phenomena. A lecture on electricity stimulated his interest. As he returned from England on his last voyage, at 79, he was still measuring the temperature of the air and of the water. Scientifically, he can be said to have been only an amateur; nature was only a secondary interest. As a physicist, however, he did postulate the increasingly significant principle of conservation of electric charge on the basis of an acceptable electrostatic theory he himself had devised. In 1753 he became the first foreigner to receive the Copley medal; in 1756 he was made a fellow of the Royal Society. In general, he was a philanthropist, involved in the founding of the University of Pennsylvania and the American Philosophical Society (1768), as well as a patriot and a writer. This undecided runaway "G" became the outstanding United States scientist of the 18th century, Benjamin Franklin.[13]

Our last group of teenagers who became scientists concerns the frustrated. We begin with No. 1, a laborer, born in 1766 at Eaglesfield in West Cumberland. His father was a poor hand-loom weaver, who made the popular gray wool coats. He lived in a thatched cottage, where the sleeping room was 15 ft $\times$ 6 ft $\times$ 6 ft. The mother was quite active; she had three children who lived, this one not being the oldest. The family (Quakers) sent the boy to a Quaker school master, who was more than the ordinary run of North Country teachers. He was found to be "not a brilliant nor quick boy," but steadfast in purpose and in power of abstract thought. At 11 he attracted the interest of a local meteorologist and instrument maker, who taught him some mathematics. At 12 he himself opened a school for children from infancy to 17. It was not a success. Accordingly, at 14 he had to work as a laborer with a plow. At 15 he decided to seek his fortune elsewhere. So with a new umbrella in hand and with his underclothes under an arm, he walked 44 miles through the Lake District to Kendal, where he worked for his brother in a school. On the side, he made barometers and thermometers. At 19 the brothers themselves took over the school. The income, however, was so small that they had to borrow money continually, even from their poor parents. They were not popular, owing largely to their uncouth manners. At 21 the young man was solving various mathematical problems for a journal and making barometric measurements of his own. He gave 12 public lectures, which had an unfavorable response. At 24 he had to collect flora and flies to make money. At 27 he decided to become a tutor in a then essentially secondary school, Manchester College, which later (1889) moved to Oxford. At 30 he found himself interested in chemistry, and three years later, accordingly, gave up his regular position. He became a private teacher, instructing very young children, in order that he might carry on private research in his small room in the back of the boarding house. Scientifically, he is noted for the first (1794) paper on color blindness (his own). In 1808 he published his famous "New System of Chemical Philosophy," for which he was made a member of the French Academy of Sciences in 1816, six years before he was selected as a fellow of the Royal Society. In 1832 he received a DCL from Oxford University. He died in 1844. This frustrated laborer No. 1 became the founder of modern atomic science, John Dalton. [14]

The bookbinder, No. 2, was born in London in 1791. His father was a blacksmith, who had to emigrate in 1791 from the country to London. When the boy was five, he lived above a coach house. The father being in poor health, the family had to go on relief when the boy was ten. The mother was a country girl; she and her husband were Sandemanians. He had an uncle who was a shoemaker and one who was a shopkeeper, another uncle who was a packer, still another a slater; his brother was a gas fitter—all good occupations, but hardly to be associated with intellectual accomplishments. The boy went to a common day school, where he learned the three R's. At 13, however, he had to take a full-time job; he became an errand boy delivering newspapers. At 14 he was apprenticed to a bookbinder, and qualified as a journeyman seven years later. Most of his education now was done by himself, mainly by reading, but including some simple experiments on chemistry and electricity, which he repeated. At 20 he was fortunate to be given some tickets for scientific lectures by the fashionable Sir Humphrey Davy. At 21 he enthusiastically wrote a letter to Sir Joseph Banks, President of the Royal Society, to ascertain if he could find employment in science—no reply. He did, however, manage to become an assistant to Davy at the Royal Institution in 1813—and its Director 20 years later. His scientific discoveries include the motor, the dynamo, and the transformer (no patents), as well as electric and magnetic properties of matter, in particular, the relationship of magnetism to light. He refused the Presidency of the Royal Society, of which he became a member at 36. At the founding of the United States National Academy of Sciences (1863), when he was 72, he was made a foreign member. He enjoyed lecturing both to adults and to juveniles. Thus this frustrated bookbinder No. 2 became the foremost experimental physicist (and chemist), Michael Faraday.[15]

Our next case was a revolutionist (No. 3) born in 1811 at

Bourg-La-Reine (France). His father was head of a boys' school after deposition as mayor. Discouraged by the loss of his social position, the father later committed suicide, when the boy was 18. His mother tutored him from time to time. At 12 he passed an examination for entrance to a secondary school, Lycée Louis-le-Grand, in Paris. At 13 in the midst of a school rebellion, he was saved from expulsion only by his absence on a critical day of action. At 16 he was put back in a rhetoric class, where he was bored; accordingly, he took mathematics for relief. Finding that the class had already reached the half-way mark of André Marie Legendre's geometry (normally a two-year course), he proceeded to read the book on his own—and finished it in two nights. He became interested in the solution of algebraic equations and later received the second prize in mathematics (his explanation was said to have been insufficient for the first). At 17 he continued to study mathematics—to the neglect of the rest of his subjects. Nevertheless, he failed the entrance examination in mathematics for the École Polytechnique. When asked by the examiner why he had not taken the customary preparatory course, he replied frankly, "I studied by myself." This sensitive young man, moreover, failed to give a complete answer to a certain question because it had been formulated so poorly. In the same year he published his first paper, which was unimportant—and unnoticed. At 18, however, he sent to Augustin-Louis Cauchy a significant article which, unfortunately, was "lost." During that year he again failed the Polytechnique examination: this time the examiner kept insisting upon a fuller explanation for a particular statement which the young man regarded as "obvious." Finally, in despair, he hit the examiner with a sponge. At 19 he published three short papers and entered the École Normale Supérieure. There he stirred up trouble with respect to the succession of Louis Philippe to the throne in place of the exiled Charles X. He was expelled and then joined the National Guard. At 20 he offered a public course in mathematics, which had to be dropped owing to the lack of attendance. Another paper was sent to the French Academy of Sciences; it was rejected by Siméon Denis Poisson, who noted the correctness of the result but complained of the briefness of the proof. He attended a banquet and gave a toast for Louis Philippe with a glass in one hand and a dagger in the other. He was arrested. The verdict was "not guilty." A preventive arrest was made later on the pretext of his wearing the uniform of the dissolved National Guard. After waiting four months for trial he was convicted and given a six-months imprisonment. At the end of the fourth month he became ill and was sent to a nursing home.

There he became interested in a girl visiting his room-mate. He fell in love with her only to find out later that she was the mistress of an absent boy friend. Upon this disclosure he gave way to an outburst and was subsequently challenged by the lover. The duel occurred on May 30, 1832. Wounded, he was left to die. The whole affair is believed to have been a frame-up. On the night before his duel, however, he wrote a scientific testament with the following pathetic notation in the margin: "I have no time." His work was neglected until 1846 and received full recognition only in 1870. Thus this frustrated revolution-ist No. 3 revolutionized all mathematics with his group theory; he himself became recognized as one of the top four mathematicians in the 19th century, Evariste Galois.[16]

A girl, No. 4, was born at Warsaw in 1867. Her father was a professor of physics and an under-inspector of schools. Later retiring from the latter position involving boarding students he had to take a smaller apartment. The mother, wellborn, pious and active, herself director of a private school, died from tuberculosis when the girl was 11. At 14 she was brilliant at the government gymnasium but neglected at home. At 16 she graduated with a gold medal. In this respect, however, she was not unlike other gifted students in her group, e.g., her brother and sister, who also received gold medals. At 17 she did some tutor-ing, but gave it up owing to the tardiness of the students and the even greater delay of their payments. Having a disdain for frivolity she adopted a severe mode of dress and had her hair cut almost to the roots. Her formal education stopped at this time owing to the policy of the Russian-controlled University of Warsaw not to admit women. At 18, therefore, she becamse a governess for the family of a lawyer who, though rich, was quite stingy, as well as vulgar and petty. At 19 she became governess for an estate administrator 60 miles north. Here she freely taught the peasants the Polish language. She fell in love with the son of the owner, but was forbidden marriage because of her inferior social position as a governess. After two years she returned to her family. And so, at 22 this young women was unhappy in love, disappointed intellectually, and hard-up materially. She had to help financially both at home and in Paris with her sister's education. At 24, however, her sister, having married, freed her of the latter obligation. At last the young wom-an herself was able to go to Paris—third class—to contin-ue her own studies. At 25 she entered the University and lived near the Sorbonne—on three francs a day. Her room had no water, no heat, no light. One day, having fainted, she was found alone in her room with only one packet of tea for food. Later she received a scholarship from a group at Warsaw; she passed first in the physics examination. At 28 she married a physicist, who was killed by a carriage 11 years later. About 30 plus she discovered new chemical elements, which she named radium and polonium. After four years, with the help of her husband, she managed to separate radium from uranium (no patent was ever sought for the novel method employed). At 36 she received the Ph.D. with the notation "très honorable." In that same year she received the Nobel prize. This frustrated girl No. 4 became the only person ever to win the Nobel prize twice for scientific achievements—Marie Sklodovsky Curie.[17]

Finally, we shall consider a high school drop-out No. 5. He was born in 1879 at Ulm on the Danube. His father owned a small electrochemical factory, which was transferred to Munich when the child was two. When the boy was 15, the father had to move his works again to Milan, owing to failure. The father was undoubtedly a free thinker; the mother, however, was of the Jewish faith. She had a sense of humor and played the piano. An uncle engineer early stimulated the boy's interest in mathematics. Although he was slow in learning to speak, he was apt with violin lessons, which he began at the age of six. He attended a primary Catholic school, but was not at all a prodigy; indeed, he continued to lack fluency in speech. At ten he entered the gymnasium for general education. Later he noted that, whereas the elementary school teachers had performed like sergeants, those in the gymnasium behaved like lieutenants; both, however, were of the military type. At 12 he began to study geometry and was so fascinated that he could not put the textbook down. At 14 he performed chamber music. At 15 he received Jewish instructions; he liked the Old Testament proverbs and ethics, but not the ritual. At this time he was far ahead of his class in mathematics, but far behind in the classics, which he complained were taught primarily for the purpose of examination. He determined to leave school supposedly on account of a nervous breakdown, but actually to be with his family in Milan. He learned, however, that the faculty had already acted and were requesting him to withdraw from school inasmuch as his fellow students were losing respect for their teachers owing to his own pronounced aversion to drill. At this time he legally renounced both his German citizenship and his Jewish religion. He failed the entrance examinations (modern languages and natural science) for the Swiss Federal Polytechnic School in Zurich. (The University there was closed to him inasmuch as he had no gymnasium diploma.) He was advised to complete his secondary education at a Cantonal school about 30 miles northwest. There, because of the independence and interest of the teachers, he experienced less aversion to school. At 17, upon graduation, he entered the Polytechnic automatically without examination. He elected the course for "training teachers in physical and mathematical subjects." Unfortunately, here he lost his interest in mathematics owing to its poor teaching by Hermann Minkowski; the physics, moreover, was out of date (he studied this subject independently). At 21 he graduated, but was unemployed for six months. Was it because of his Jewish background or his colleagues' jealousy that he received no assistantship? Neither could he obtain a teaching position in a secondary school. He did, however, manage to become a tutor for two students in a grammar school, but was fired when he tried to become the boys' sole teacher. Eventually he had to resort to a job in the Berne Patent Office. As to his scientific accomplishments, at the age of 26 while at the Patent Office, he wrote three papers, any one of which would have won him world renown. The one on the photoelectric effect was based upon the recent quantum theory and was the primary basis later for his receiving the Nobel prize in physics. The second, on Brownian movements, was subsequently used for direct experimental confirmation of the century-old atomic theory. The third, quite original, was on relativity. He became professor at the Universities of Zurich and of Prague, at the Prussian Academy of Sciences and at the Institute for Advanced Study in Princeton. He was a pacifist and a Zionist; he is frequently remembered in the United States for a letter he wrote recommending the construction of an atomic bomb to President Franklin D. Roosevelt. Thus this frustrated high school drop-out No. 5 revolutionized all physics in the 20th century—Albert Einstein.[18]

Each teenager is concerned today with the question: How far will I get? He is told that his progress will depend upon his heredity and his environment. Probably it is fortunate that our personal heredity is so largely unknown and that very few of us ever reach its limits. It is profitable, therefore, to consider the experiences of other individuals[19] such as the advantaged, the undecided, and the frustrated, who, although they may have lived in a different social scene, as human beings, nevertheless, performed on a similar stage. Each one was confronted with personal difficulties and indecisions, which he overcame by energetic perseverance and enthusiastic interest. Such lives can be object lessons to us all. They suggest clues for solving our own problems. A student of mine once looked questioningly when I handed out some homework and said indignantly, "Who do you think I am, Einstein?" I replied, "No, Einstein would probably not be taking this physics course the fourth time." It is not always the

obviously bright who go the farthest. You recall the story
of the race between the tortoise and the hare. On that
beautiful spring day if you had been betting on the basis
of heredity and environment you would undoubtedly have
selected the hare—and lost. I, myself, had a similar experience some years ago. Getting into my car at night to
drive some 200 miles from Washington to Durham, N. C.,
I suddenly realized that my headlights shone only 200
feet. How could I go the necessary 200 miles? Going those
200 feet, however, I noticed that the lights still shone
another 200 feet. Thus by always going the distance that I
could see, I ended my journey at a point that had been
far beyond my vision. It is said of an alpine climber that
he was last seen going forward.

As a graduate student I had occasion to hear a professor of eugenics speak on the subject, "Who Shall Inherit
the Earth?" He argued that obviously the sons and daughters of the learned, the well-to-do, the leaders of society,
et al. would inherit the earth. When he had finished, an
editor of one of the New Haven newspapers said, "Professor, you are correct. Those are the ones that will inherit
the earth. But allow me to tell you who will take it away
from them: the sons and daughters of the Irish charwom-
en on State Street." (That man was the father of
Thornton Wilder.)

Some years ago I heard a story about an Indian, which
seems to summarize succinctly what I have been trying to
say about motivation. A secondary school teacher of music happened to visit a small town some hundreds of miles
from Mexico City to give a concert. Seeing a 10-year old
barefoot Indian boy, he asked if the boy would help pump
the organ at the evening concert. The teacher became
interested in the boy and inquired the next day if the boy
could go along with him as a personal valet. The mother
and father were thrilled: it would be a rise in the social
scale, for the boy could now wear shoes (furthermore, the
family was again "expecting"—number 17). The boy was
discovered to have some musical talent, and later, at 19,
performed in the presence of President Parferia Diaz. The
President, who had been told the story, offered the young
man a scholarship to study music at the Paris Conservatory of Music; the Conservatory, however, turned him
down owing to his "advanced" age. Nevertheless, the boy
became a concert violinist, Director of the Symphony of
Mexico City, the Dean of the University's School of Music. Records of his analysis of the songs of birds are to be
found in the Library of Congress. His son graduated from
Harvard with a Ph.D. in soil mechanics, and became
President of the University of Mexico—he told me the
whole story himself.

May I remind you of a perennial challenge, as stated some years ago by James B. Conant: "To find and educate the gifted youth is essential for the welfare of the country, we cannot afford to leave undeveloped the greatest resource of the country."

[1] "Physical Science Aptitude and Attitudes Toward Occupations," Lafayette, Ind., Purdue Opinion Panel Research Report 45 (July 1956).

[2] Dorothy G. Rodgers, "An Analysis of Attitudes Toward Science," Lafayette, Ind., Purdue Opinion Panel Research Report 58–41 (May 1958).

[3] Emily Post, *Children Are People* (Funk and Wagnalls Company, New York, 1940).

[4] Morris Bishop, *Pascal* (Williams and Wilkins, Baltimore, 1936).

[5] Alexander Wood and Frank Oldham, *Thomas Young* (Cambridge University Press, New York, 1954).

[6] Lewis Campbell and Wm. Garnett, *The Life of James Clerk Maxwell* with selections from his correspondence and occasional writings, New ed. rev. (Macmillan and Company, Ltd. London, 1884); Richard Tetley Glazebrook, *James Clerk Maxwell and Modern Physics* (The Macmillan Company, New York, 1896).

[7] Max Caspar, *Kepler*, translated and edited by C. D. Hellman (Abelard Schuman, London, 1959).

[8] J. J. Fahie, *Galileo* (John Murray, London, 1903); Laura Fermi and Gilberto Bernadini, *Galileo and the Scientific Revolution* (Basic Books, Inc., New York, 1961); Raymond J. Seeger, *Galileo Galilei, His Life and His Works* (Pergamon Press, Inc., Oxford, England, 1966).

[9] E. N. DaC. Andrade, *Isaac Newton* (Chanticlear, New York, 1950); Louis Trenchard More, *Isaac Newton* (Dover Publications, Inc., New York, 1934, 1962).

[10] Angus Armitage, *William Herschel* (Thomas Nelson and Sons, London, 1962).

[11] Sanborn C. Brown, *Benjamin Thompson—Count Rumford* (Pergamon Press, Inc., Oxford England 1967).

[12] Gavin De Beer, *Charles Darwin* (Doubleday and Company, Inc., Garden City, 1964); Julian Huxley and H. B. D. Kettlewell, *Charles Darwin and His World* (The Viking Press, New York, 1965).

[13] *The Autobiography of Benjamin Franklin* (Pocket Books, Inc., New York, 1952); Carl Van Doren, *Benjamin Franklin* (The Viking Press, New York, 1938, 1965).

[14] Sir Henry E. Rascal, *John Dalton and the Rise of Modern Chemistry* (Cassell, London, 1901).

[15] John Tyndall, *Faraday as Discoverer* (Apollo, New York, 1961).

[16] Leopold Infeld, *Whom the Gods Love* (McGraw-Hill Book Company, New York, 1948).

[17] Eve Curie, *Madame Curie*, translated by V. Sheehan (Doubleday and Company, Inc., Garden City, 1943).

[18] Philipp Frank, *Einstein*, translated by G. Rosen, edited and revised by S. Kusaka (Alfred A. Knopf, Inc., New York, 1947).

[19] F. Sherwood Taylor, *An Illustrated History of Science* (William Heinemann, London, 1955).

Editor's Note: This article was presented as an invited address at the annual meeting of Association for the Education of Teachers in Science—National Science Supervisors Association 1 April, 1968, Washington, D. C.

Chapter Two

TWO MODES OF KNOWING

In his recent book, On Knowing, Essays for the Left Hand (Atheneum, New York, 1965), Jerome Bruner uses the symbolism of right and left hand to contrast action, logic, and reason with dreams, intuition, and sentiment. Science is traditionally associated with the right hand; art, poetry, and religion with the left. Yet, how accurate is this stereotype? In Chapter 1 Arthur Koestler found it necessary to represent scientists through a model that had three contrasting elements. Look for "right-handed" and "left-handed" knowing on the part of the individual scientists described in Chapter 1 and in your textbook. What is your own preference?

Now we turn to science itself. Merle A. Tuve accepts the different approaches of physics and the humanities in his article, reprinted here. He advances the view, however, that the two supplement one another rather than being in conflict. To clarify the relationship further, he proposes an analogy with the wave and particle aspects in the dual nature of matter and of light, which are described in Chapter 8 of your text.

How do poets look at physics? Three answers to this question are illustrated in the selections by contemporary poets. The short poem by A. R. Ammons paints a verbal picture of a phenomenon also of interest to the physicist, but he says a great deal more than Figure 5.4 in your text can communicate. David Wagoner draws an analogy between the concept of inertia as defined in Newton's first law of motion and certain aspects of the human experience. And yet, don't the differences seem to be more important than the similarities? Physics itself is the subject matter of Josephine Miles, who appears to have been influenced by her scientific colleagues at the University of California. Which of the three answers appeals to you? Are there still others?

The last two selections in this chapter, by Warren Weaver and by Daisaku Ikeda, attempt to relate science and religion on the intellectual plane, the domain of the right hand. Even though they reflect two very different religious traditions, both articles emphasize the connections of religion with the theoretical or model-building aspects of science. Warren Weaver, the scientist, finds in God reassurance with regard to the tentative and probabilistic nature of scientific explanations and predictions. For Daisaku Ikeda, the religious leader, ancient Buddhist doctrines foreshadowed modern scientific theories.

Does this mean that faith and religion are needed to legitimize science? What is your experience and your opinion?

About the Contributors

MERLE A. TUVE (1901-   ) is a Distinguished Service Member of the Carnegie Institution after serving as director of its Department of Terrestrial Magnetism for twenty years. Dr. Tuve has won many awards and prizes in recognition of his work in geophysics and radio propagation.

A. R. AMMONS (1926-   ) is Professor of English at Cornell University. He holds a bachelor's degree in general science and did graduate work in English with Josephine Miles, to whom his first book, Ommateum, is dedicated. For several years Mr. Ammons was an executive in a glass factory. In 1961 he participated in the Bread Loaf Writer's Conference and he has recently served as poetry editor of The Nation.

DAVID WAGONER (1926-   ) is Professor of English at the University of Washington, Seattle, and editor of Poetry Northwest. His latest collection of poems is New and Selected Poems (Indiana University Press, 1969), for which he won an award from the National Council on the Arts. Mr. Wagoner is also a novelist and has written five novels, the most recent of which is Baby, Come on Inside (Farrar, Straus, and Giroux, 1968).

JOSEPHINE MILES (1911-   ) is Professor of English at the University of California, Berkeley, where she has served on the faculty since 1941. She is a poet, literary critic, and teacher. Among Miss Miles' many publications are the collection of criticisms Style and Proportion (Little, Brown, 1967) and two college texts. She has won numerous awards for her writings.

WARREN WEAVER (1894-   ) is now a consultant to the Alfred P. Sloan Foundation after many years as director of the Rockefeller Foundation. Though he originally trained for mathematics research and teaching and was active in those areas, Dr. Weaver's most significant contributions to society are the foundation-supported programs carried out under his leadership. These include agricultural improvement and education leading to the large-scale use of hybrid corn in Mexico, and the research in molecular biology which has revoluntionized man's conception of life. Dr. Weaver is also an outstanding science writer, winning the UNESCO's Kalinga prize in 1965. His books Alice in Many Tongues and Lady Luck are recommended for laymen.

DAISAKU IKEDA (1928-   ) is the president of Sokagakkai, a dynamic Buddhist group with rapidly growing membership in Japan. Mr. Ikeda has published many writings on Buddhism in an effort to highlight the teachings of that religion as they are relevant to modern social problems. His approach deviates in some respects from that of many other Buddhist sects.

PHYSICS AND THE HUMANITIES— THE VERIFICATION OF COMPLEMENTARITY

Remarks on receiving the Third Cosmos Club Award, in Washington, D. C., May 9, 1966.

Merle A. Tuve

... If we survey the actual life-matrix in which our thinking is done, and by which it is obviously conditioned, we find most emphatically that life for each of us is much wider than our particular professional interest and competence. In addition to the vague, sometimes humdrum, and sometimes ecstatic sense of being which fills most of our waking hours, each of us has been much involved with specialized areas of thought and action which are classified as belonging to other men's fields of special interest or professional competence.

In the course of a great variety of experiences during the past 30 years I, too, have found myself concerned with ideas and actions in specialized areas quite outside of my own field of physics. Among these topics there have been several which seemed to me perhaps appropriate for my talk here tonight . . . but after pondering . . . questions of general public significance, and trying to see by what right of thoughfulness —or arrogance—I might consider myself qualified to discuss them in the distinguished company of experts which is the Cosmos Club, I finally concluded that I should speak out as a physicist, not as a social commentator. . . . It has seemed best . . ., however, for me to examine the ways in which physics makes its most basic contributions to modern society, contributions to the universal search by every man, when he is not totally preoccupied by hunger or by some immediate threat of catastrophe, for fullness and meaning in his life, the search for richness of experience, set in categories of value—in short, the search for significance in his life.

Most of us, at first encounter with questions about the contributions of physics, think at once of atomic power or atomic bombs ("to preserve the peace"); then we think of the marvels of electronic communication, then of air transportation and space travel, then perhaps of computers and the automation of industry. Then the mood darkens and we think of the troubles of human displacement and adjustment, and the accelerating pace of life which dominates and threatens even us, as the world's most favored few, and we wonder if any gift of modern science yields to human beings anything but power and the impersonal domination of more and more complex technological systems, seemingly with uncontrollable destinies of their own, as mankind cowers, and the individual feels more and more lost and estranged.

Oddly enough, these are not problems of science—they are all *problems for the humanities* to resolve to their ultimate roots in human hopes and needs, and then to guide the social sciences and governments and peoples toward compromise and resolution, and finally, to lift men's eyes toward higher levels of brotherhood and richness of personal experience. We all know this, and we all recognize that these problems of the innate plus the cultural characteristics of mankind, so universal in some ways, and yet expressed in such contradictory ways in the various groupings of cultures and of nations, are commandingly urgent. They are much more urgent for us to resolve constructively than our needs for more knowledge concerning the extremely rare and fantastically short-lived particles that we knock out of atomic nuclei using the huge accelerators of high-energy physics, or our needs for the satisfaction of our competitive drives and the possible answering of a very few questions of very specialized nature by our present vast Federal efforts for, and commitments to, of all startling things, the man on the moon. This particular error of public emphasis seems absurd, on the face of it, when we consider the enormously destructive potential of the emphatic questions of human maladjustment and disagreement now conspicuous throughout the world, and the new dimensions of social power and individual weakness. Yet, so to speak, we fiddle while Rome burns. This kind of distortion in our public and private lives is one example of the ways in which technology, unless guided by philosophy, can threaten and defeat the good life.

I am known to be one of the early and steadfast critics inside the general fraternity of the physical sciences—in "big science" circles, if you favor that epithet—opposing the general notion that just because we think a new big thing can be done we must now do it, and, correspondingly, our government must pay the bill. I think that in science, as in all other human affairs, a touch of austerity is a necessity for health, and our powers for discrimination and relevance must be fully exercised, or we will surely spend our efforts on conspicuous or gaudy projects, not on those of basic concern.

So I sat in my study in the quiet of the night and asked myself these questions: What are the really basic contributions of modern physics to the areas of the good life, to the full and significant experience of living? What specific elements of importance does physics yield for individual persons, for you and me and for whole societies, among them our own Western cultural groups? Has physics made contributions which may also be relevant and important to those individuals who have inherited the magnificently different traditions of the Far East, and view life from a background of Oriental philosophy? Has it value and meaning even for the fragmented groups now seeking to leap from narrow tribal confines to a full participation in world society? What do I really think are the outstanding bequests from physics to the immediate future of human satisfaction and fulfillment? Are any of these contributions possessed of such sure inherent qualities of permanence that they will continue to contribute to the lives of men, and to their sense of wholeness and unity with Nature and with God, over the uncertain long and upward reaches of the future?

This whole area of personal contemplation and assessment—indeed the entire development of the mental and emotional content of significant living—the ways by which we enjoy and enlarge the richness of human experience—we seem to have allocated to the general area referred to as the humanities and the fine arts. As I sat there in my study, my thinking slowly came to a focus on one very important contribution of modern physics in this intimate field of personal awareness and significance. This contribution is called *complementarity*.

As a physicist I am an experimentalist, but as such I have always been much concerned, of course, with interpretation and theory. Theoretical physics has a language of its own, almost impenetrable, but, like the doorkeeper at the house of the Lord, I have long stood at the door of theoretical physics with my ears open.

There is no necessity for me to paint mental pictures showing the vast scope of experimental physics, ranging from distant quasi-stellar objects of astrophysics to the mesons and strange particles of nuclei and high energy physics, and even farther to the deceptive simplicities of the DNA and RNA of biophysics. Instead, I shall simply draw your attention to one basic clearcut result of modern theoretical physics which has, in my judgment, immeasurably wide and deep significance for every man.

The verified necessity for us to accept two very different views of natural events, mutually irreducible one to the other, a necessity which has been given the name of *comple-*

mentarity, is one great gift of physics in our epoch to the thinking of all humanity. In what follows I shall endeavor to indicate something of the content of this sober remark.

I can express tonight only feebly what has been much better said by others, namely, that the ultimate effect on human life and endeavor and satisfaction of this precious fragment of new understanding, based irrevocably on simple experiments of modern physics, may be far more profound than any of the technological offshoots of atomic energy or electronics or automation which also trace their genealogy to the curiosity and the research studies of the physicists. A number of my friends have expounded the nature of belief or conviction, the vast range of our awareness and the limited fraction of it which is covered by physics, and have gone on to expound the pertinence of the concepts of complementarity and the uncertainty principle in relation to man's own personal acceptance of the dichotomies or antinomies of human awareness. Among these friends are Warren Weaver and Robert Oppenheimer, to whom I owe, without their knowing it, much of the courage needed to make this address.

Over the years I have been privileged to know personally many or most of the giants of this splendid modern enterprise of human understanding, the demonstration of the intrinsic fact of complementarity in the physical world. They were few in number. Perhaps they were fewer than a dozen, all associated intimately and frequently with Niels Bohr, of Copenhagen, who made many visits to the U.S.A. and shared himself with all of us. I personally was privileged to share only in a very small way in the yeast of these developments as they happened, mostly as a listener, although later my colleagues and I shared the related pleasure of making, here in Chevy Chase, the first quantitative measurements of proton-proton and proton-neutron force interactions—the first measurements of the binding force which, by $E = mc^2$, makes atomic nuclei weigh less than the sum of their constituents. (Incidentally,

I might remark that I then proposed in a letter to *The Physical Review* that because of this weight loss the nuclear forces might be called the force of levity, which is surely as fundamental to all Nature as the force of gravity! To this day, however, my sober-minded scientific friends have elected to disdain that gay remark, much as though I had broken out with applause in church.)

The studies of nuclear physics using high-energy particles from accelerators really came after the great ferment of struggle and insight brought about in atomic physics by the nuclear atomic model of Rutherford, the Bohr atom model, and the mathematical formulations of Bohr, Pauli, Heisenberg, Schrödinger, deBroglie, Dirac, and a few others in the period 1911 to 1930. The most intense period was perhaps 1924 to 1928. By these men an abrupt and permanent mutation in human thinking was completed about thirty years ago, and was expressed in two or three basic ideas or statements referred to as the principles of correspondence, of complementarity and of indeterminance.

I shall indicate only very briefly the nature of these ideas. The Bohr atom of 1912, with its quantum transitions between steady states, quite irrational and absurd from the viewpoint of classical Newtonian mechanics and Maxwellian electrodynamics, proved experimentally and empirically correct beyond all possibility of doubting, yet defied all possibility of understanding in terms of previous dynamics. Various mathematical formalisms were devised which simply "described" atomic states and transitions, but the same arbitrary avoidance of detailed processes, for example, descriptions of the actual *process* of transition, were inherent in all these formulations.

During this same period, around 1925, it was demonstrated by G. P. Thomson in England and by Davisson and Germer at the Bell Laboratories in New York that electrons undeniably behaved as waves in certain types of experiments which asked typical questions formulated in terms of wave ideas, such as the positions of interference maxima and minima, yet the arrival of the electrons at these wave positions was never possible to show as waves but always, and only, as separate and discrete particles. After some years of puzzling as to the nature of this wave description, it was realized that the most nearly correct statement was that they were waves of probability. The wave does not predict the actual appearance of an electron as a particle, but only the relative probability, given large numbers of electrons, for the appearance of elec- trons at the positions indicated by the intensity of the computed waves for these positions. . . . This *complementarity* of the electron both as a wave and a particle is a basic, inescapable dichotomy, on a par with the irrational dynamics of Bohr's nonradiating stationary atomic states and the lack of any conceptual notion which might describe the actual process of the transition of the Bohr atom from one stationary state to another.

Other detailed analyses, especially by conceptual experiments (in contrast to laboratory experiments), led to the inescapable realization that the process of observing or measuring, when we deal with space and mass and energy of atomic dimensions, affects the result of the measurement itself, much as wave questions give wave answers and particle questions give particle answers, when we examine the behavior of electrons. This effect of the observer was most graphically expressed by saying that if we seek to know the exact position of an atomic event, such as a collision, we must be content to know nothing about the momentums involved, or if we ask in detail about momentum exchange, we must be content to remain ignorant of exactly where the event took place. This kind of technical statement is one expression of the basic fact that we cannot learn or predict atomic processes in full detail, which is Heisenberg's Principle of *Indeterminance*.

For ordinary massive bodies the old Newtonian dynamics and Maxwellian electrodynamics were not altered by any of these discoveries or conceptual experiments, even out to extremes of large size and mass and energy. This recognition was expressed by the word *correspondence,* which affirmed the fact that all atomic equations must reduce to the classical equations (as earlier revised in certain ways by Einstein and by Planck) when aggregates of matter, large compared to single atoms, are considered.

By 1931 this radical change in our basic philosophical approach to the laws of Nature was complete.

Now for 35 years man has slowly begun to recognize the implications of this discovery that there are immutable and inescapable dichotomies in his views of the same objects in the physical world, depending on the kinds of observations he makes, the kinds of questions he asks.

The shockingly new and far-reaching discovery, for one's everyday philosophy of living, is that the most abstruse calculations and knowledge of the most detailed and unexpected phenomena of the atomic world which we observe and measure have led unequivocally to a profound verification of an age-old inner awareness which is the most immediate and commonplace philosophical experience of every individual person.

From the time we were adolescents we have all been troubled by the obvious fact that there are different modes of examination of our individual experiences in living, alternate modes roughly characterized by the differences between the sciences and the humanities, but on an internal basis, which we experience as the contrast of *rational* factors with *emotional* factors. And now we know, from the most detailed and profound examination of the outside, objective, physical world around us, that the true answers about reality are intrinsically determined by the way we frame the questions. Furthermore, we find that these answers can even be *logically incompatible* with each other in very fundamental ways, characterized by the nature of the observations or the assessments we make, which are chosen to fit the nature of the particular questions to which we address our attention.

When we ask whether electrons are waves or particles, or whether ordinary light is emitted and travels as waves or as particles, we learn from both theory and experiment that we are compelled by Nature to say, with clear inner contradiction, *both,* but *not* both at once, only one or the other, depending on the *question* you ask—and this you *must ask first.* [This is like the definition of infinity: you name a number first, then I can name a larger number. This describes a process, not a finite stationary state which can be localized and identified.] We have the same experiences with categories of our own personal thought or attention; blood may rush to our cheeks and our hearts may pound with embarrassment, but this is not equivalent to, or interchangeable in any way, with our intense awareness of the moment or our thoughts about the faux pas which precipitated the blush. The motivation, and the initiative for retrieving the situation, again are not in the same category with the biochemical processes which activate our muscles as we retreat from the encounter, however simultaneous in real time these parallel realities may occur or exist.

In biophysics and biochemistry we have been pressing forward in the search for the "physical basis" of various self-propagating or self-activating expressions of living matter, all the way from genetics and the biological coding of DNA and RNA for the synthesis of proteins and the determination of enzymatic behavior, to the roles of RNA and protein synthesis in the formation and retention of memory, which is the acquisition of knowledge by an individual. Half of the research effort of our own Carnegie Laboratory in Chevy Chase for the past 20 years has been devoted to these basic aspects of biophysics. Yet, no man can claim that a description of the physical processes which are simultaneous with a thought or an emotion are fully and identically the *same* as that thought or emotion.

This is not the kind of resolution a physical scientist can seek for the mysteries of life and thought and feeling. Far more acceptable is the intuitive answer men have always given, from the most primitive savage to the most sophisticated intellectual, that things of the mind and spirit of man have a reality all their own, parallel with, but not the same as, our participation in the physical world. Here again, the answers you find are basically determined by the questions you ask. This is an age-old expression of a similar kind of *complementarity,* sensed and expressed and never "resolved." Modern physics, unexpectedly, and in the experimental world of the laboratory, has provided an unequivocal demonstration of the finite or limited nature of our human possibilities for "understanding" the world in which, unasked and without choice, we find ourselves. The message is clear and direct, and leads both to confidence, to faith, and to humility. There *are* different ways, conflicting ways too, in their very essence, by which we view the world around and in us, but the true answers will be framed in terms which fit the questions which we ask.

This verification of complementarity, however, is decidedly not a license to think whatever you please, or to believe to be true whatever ideas may appeal to you. Atomic physics has given an unequivocal demonstration that two sets of ideas, as contradictory as diffusely spread waves and particles concentrated at singular points, are both true descriptions of the material aspect of Nature, complementary views of reality, each of which is necessary in its own turn to answer special questions men have learned to ask. But this does not mean that reality or truth is so unlimited in scope, so multiple-valued or various in its facets, that every fancied description is as valid as any other. Instead, it comprises an unexpected but very specific proof of the limits of the area of authoritative competence of those individuals whose excessive confidence in the rational or scientific approach to the mysteries of Nature and comprehension leads them to discard all the other aspects of awareness and life experience. At the same time, it constitutes an invitation and a challenge to esthetics and the humanities to seek new and different expressions of the spirit of man. It is a fresh encouragement to those among us who are continuing the long search for criteria of validity in those vast realms of ideas and personal awareness which cannot be reduced to axioms and logic and measurement, which defy classification systems, which often elude us and largely disappear when we subject them to the rigidities of language.

Truth and beauty in science and technology comprise only a minor fraction of the total of truth and beauty in human awareness. It is a major step, of course, and vastly comforting, for physics to reach a reasonably comprehensive description of the small and large examples of material things which surround us in this physical universe, and to have these descriptions substantiated by more and more observations. Complementarity and correspondence principles indicate the finite limitations of human concepts, and indeterminance expresses an appropriately corresponding humility. It has taken about 2,500 years of interest and puzzlement and thinking to reach this tentatively satisfactory, though highly circumscribed and limited, world view.

There is something quite poetic about the fact that the capstone of all this achievement of ideas in the realm of the rational aspects of our thought-processes, viewing primarily those material items around us which are not endowed with the special property we call living or life, is this demonstration of the finite or limited scope of our idea processes. We have wondered in physics about the kinds of research which will be interesting, or even possible, 50 or 100 years hence. Most of the basic laws seem known, and the simple things, the one-man accomplishments, seem to have been already studied and done.

This state of affairs is in great contrast, I would guess, to the situation in the humanities and esthetics and the fine arts. Perhaps there are the equivalents of Newton's Laws and

Galileo's telescope and the conservation of energy in these other areas of human contemplation, but if so they seem pretty much obscured by the overlaid accretions of many overzealous generations in each of many different cultures. The whole concept of such equivalents I would myself expect to be erroneous. Surely we have, however, some key concepts for organizing our thoughts and feelings in these areas, concepts such as justice, love, freedom, beauty, honor, and their opposites, and we have systems of thinking, full of similarities and agreements, as outlined in various religions and philosophies, which most certainly are not all wrong or pointless. In fact, the generally accepted public philosophy largely governs the actions of a people, and difficulties have arisen among men and societies whenever the mechanisms of society became too visibly different from the ideas held by that society in the realm of the humanities and personal convictions.

The chief point to be made, perhaps, is that if physics is running out of interesting problems, the situation is quite the opposite in these other areas of personal awareness and ultimate concern. A subsidiary point, which is the theme of this talk, is that physics has put forward, as a permanent major feature of man's knowledge of matter and space, a special disclaimer, a special denial of the uniqueness and all-encompassing validity of the rational and material ideas of science, vaunted by some philosophers and helplessly accepted by so many others less well equipped to judge. Modern physics has given to all a specific demonstration that more than one set of ideas must be used if our finite minds are to view and comprehend in some increasing measure of completeness the fabulous complexity and beauty of the awareness which is the central mystery of life for each of us.

Physics works in four orthogonal or independent dimensions, *x, y, z,* and *time.* Because time flows only in one direction, so to speak, it has special properties, expressed in part by relativity. For two or three decades I have tried to indicate the nature of these other aspects of human awareness by saying that they are components extending in a fifth dimension, the esthetic or spiritual dimension, which is orthogonal to x, y, z, and t, and equally real. Here is the locus of such nonmetrical realities as love, beauty, justice, freedom, and honor, and this dimension is as immediate to our perception and as definite a ground for our being as the x, y plane of this floor beneath our feet. This statement is of course only a mode of description, but most scientific men seem to forget that all information is first subjective and immediate, and its objectivity develops only after it has been processed by various operations.

There is nothing original in my remarks to you this evening; all these things have been better said by others. I simply have felt that, as a scientific member of the Cosmos Club I should speak out to those others among us whose activities and competences lie more in the direction of the humanities and the social sciences and the fine arts.

Actually, as you can sense from what I have said, my life experience and my warm conviction is that physics and the natural sciences, especially as they have flowered and enlarged their concerns with human interests during the past half century, are a part of, and participate with vigor in, the humanities and the fine arts, and through philosophy and by their technological off-spring, forcefully condition the social sciences as well. There are materials enough in this one remark for several more quiet discussions among us, but tonight I have wanted only to focus your minds on the liberating quality, for all philosophers and humanists, of the one single contribution made by modern physics in formulating and verifying *complementarity* as a basic property, an intrinsic, inescapable attribute, of the real physical world, insofar as the finite mind of man is ever to be allowed to view reality.

The best expression that I know of these and similar thoughts was given by Robert Oppenheimer in 1953 in a series of Christmas lectures over the BBC in London called the Reith

Lectures, published by Simon and Schuster under the title *Science and the Common Understanding.* I, personally, have long felt deeply grateful to "Oppie" for these and many other insights, ranging all the way from physics to philosophy, and I will share with you a few paragraphs from his beautiful presentation for the BBC. He speaks of the physico-chemical description of living forms and the question whether, despite the electron microscope and radioactive tracers, this kind of description can in the nature of things ever be complete. I quote:

> Analogous questions appear much sharper, and their answer more uncertain, when we think of the phenomena of consciousness; and, despite all the progress that has been made in the physiology of the sense organs and of the brain, despite our increasing knowledge of these intricate marvels, both as to their structure and their functioning, it seems rather unlikely that we shall be able to describe in physico-chemical terms the physiological phenomena which accompany a conscious thought, or sentiment, or will. Today the outcome is uncertain. Whatever the outcome, we know that, should an understanding of the physical correlate of elements of consciousness indeed be available, it will not itself *be* the appropriate description for the thinking man himself, for the clarification of his thoughts, the resolution of his will, or the delight of his eye and mind at works of beauty. Indeed, an understanding of the complementary nature of conscious life and its physical interpretation appears to me a lasting element in human understanding and a proper formulation of the historic views called psychophysical parallelism.
>
> For within conscious life, and in its relations with the description of the physical world, there are again many examples. There is the relation between the cognitive and the affective sides of our lives, between knowledge or analysis and emotion or feeling. There is the relation between the esthetic and the heroic, between feeling and that precursor and definer of action, the ethical commitment; there is the classical relation between the analysis of one's self, the determination of one's

motives and purposes, and that freedom of choice, that freedom of decision and action, which are complementary to it. . . .

To be touched with awe, or humor, to be moved by beauty, to make a commitment or a determination, to understand some truth — these are complementary modes of the human spirit. All of them are part of man's spiritual life. None can replace the others, and where one is called for the others are in abeyance.

The wealth and variety of physics itself, the greater wealth and variety of the natural sciences taken as a whole, the more familiar, yet still strange and far wider wealth of the life of the human spirit, enriched by complementarity, not at once compatible ways, irreducible one to the other, have a greater harmony. They are the elements of man's sorrow and his splendor, his frailty and his power, his death, his passing, and his undying deeds.

Now I invite all of you as fellow humanists to make the most of this new and fresh awareness, coming unexpectedly, perhaps, but in most explicit fashion, from the field we know as physics, this new verification, confirming our intuition of complementarity, inviting and urging us to use whatever modes of questioning and of thought and expression are best fitted to evoke and to express the spirit of man, as he perceives the awesome beauty of the world of which he is a finite part, yet, also, inexpressibly, of infinite extension.

OMMATEUM, 4

by A. R. Ammons

I broke a sheaf of light
  from a sunbeam
that was slipping through thunderheads
drawing a last vintage from the hills
O golden sheaf I said
and throwing it on my shoulder
brought it home to the corner
  O very pretty light I said
  and went out to my chores
The cow lowed from the pasture and I answered
yes I am late
already the evening star
The pigs heard me coming and squealed
From the stables a neigh reminded me
Yes I am late having forgot
I have been out to the sunbeam
and broken a sheaf of gold
  Returning to my corner
I sat by the fire and with the sheaf of light
that shone through the night
and was hardly gone when morning came

THE FIRST LAW OF MOTION

by David Wagoner

"Every body perseveres in its state of rest, or of uniform motion in a right line, unless it is compelled to change that state by forces impressed thereon." Isaac Newton, Principia Mathematica

Staying strictly in line and going
Along with a gag or swinging
Far out and back or simply wheeling
Into the home-stretch again and again,
Not shoving or stalling, but coasting
And playing it smooth, pretending
To make light of it, you can seem
To be keeping it up forever, needing
Little or nothing but your own
Dead weight to meet
The demands of momentum,
But there's no way out of touching
Something or being touched, and like it
Or not, you're going to be
Slowing down because turning
A corner means coming to a dead
Halt, however slight, to change direction,
And your impulse to get moving
Again may never move you, so keeping time
Is as inhuman as the strict first law
Of motion, and going off on your own
On some lopsided jagged course
For which there's no equation, some unbecoming
Switchbacked crossfooted trek in a maze
Of your own invention, some dying
Fall no star could fix, is a state of being
Human at least, and so, at last, is stopping.

PHYSICS

by Josephine Miles

The mean life of a free neutron, does it exist
In its own moving frame a quarter hour?
In decibel, gram, ohm, slug, volt, watt, does it exist?
Not objects answer us, not the hand or eye,
But particles out of sight.

Does a book in equilibrium on a shelf
Compose its powers? It upsets my mind.
Turbulent flow, function of force times distance,
The sledge with a steel head
Is energy transformed.

Illusion boils a water into cold,
A speed of pulse slows by chronometer,
A camera's iris diaphragm opens wide
To faint light. Lenses
Render to us figures equivocal.

Sight in its vacuum, sound in its medium strike so aslant
The thunder relishes a laggard roll,
And as long waves of low pitched sounds bend around corners,
Building-corners cut off a high wrought bell,
To set its nodes and loops vibrating symmetrically at the surface.

Neutron transformed, neutron become again,
In glass and silk, tracing a straight world line,
Exists in its gravitational electromagnetic fields
Trembling, though beyond sight,
Initially at rest.

✳ Can a Scientist Believe in God? [1]

What is science? It is the activity whereby man gains understanding and control of nature. It is practiced professionally and intensely by a few, but practiced to some degree by everyone. It proceeds by observing and experimenting, by constructing theories and testing them; by discarding theories that do not check with facts, and by improving good theories into better ones. It never is perfect, never absolute, never final. But it is useful, and it improves.

Not every scientist would accept this definition. Almost every scientist would want to change it a little, and a few would change it a lot. But, by and large, a scientist is ready to define science. He doesn't feel the need (as he would feel in trying to define religion) to qualify his statement by saying: "This is *my* kind of science; this is what science means to *me*."

What is religion? Religion is a highly personal affair. I can only tell you what the word means to me. Religion, to me, has two main aspects. It is, first, a guide to conduct. Second, it is the theory of the moral meaning of our existence. Do not be surprised that this definition of religion has a practical aspect that touches every act of every day, and a more "intellectual" aspect that comes into play relatively seldom. This double answer is to be expected from a scientist, as we shall see. And scientists are precisely the kind of people who should not be surprised if these two aspects are not "consistent" with each other.

Science tries to answer the question: "How?" How do cells act in the body? How do you design an airplane that will fly faster than sound? How is a molecule of insulin constructed?

Religion, by contrast, tries to answer the question: "Why?"

Why was man created? Why ought I to tell the truth? Why must there be sorrow or pain or death?

Science attempts to analyze how things and people and animals behave; it has no concern whether this behavior is good or bad, is purposeful or not. But religion is precisely the quest for such answers: whether an act is right or wrong, good or bad, and why.

How do you define God? Some regard God in very human terms, as a father who is kind but nevertheless subject to spells of wrath. Some assign to God a lot of other human qualities (love, anger, sympathy, knowledge, etc.), but expand these qualities beyond the possibilities of mankind (limitless love, infinite wisdom, total knowledge, etc.). Still others take a mystical attitude toward the concept of God: God is a spirit, and it would not be useful or possible to describe God in any other way.

I am sure that each of these ideas has well served different persons at different times. The difficulty I find with the three conceptions of God just summarized is not that they are vague; not that they depend upon faith rather than reason; not even that they may involve contradiction. I think that vagueness is sometimes not only inevitable but even desirable; that faith, in certain realms of experience, is more powerful than logic. And scientists accept such metaphysical contradictions more readily than most people think.

My difficulty with the views of God sketched above is simply that, although they bring comfort on the emotional plane, they do not seem to bring satisfaction on the intellectual plane. When I take any such idea of God and try to work with it mentally—try to clarify it or think it through—I find myself getting confused or embarrassed, using words with which I am not fundamentally content, words which cover up difficulties rather than explain them. It therefore gratifies me to use additional ways of thinking about God—ways which seem to me intellectually satisfying and consistent with the thinking I try to do along other lines—scientific or not. Indeed, it is these additional ways which very directly relate to scientific thinking and scientific theories. Let me llustrate this.

When I am troubled or afraid, when I am deeply concerned for those I love, when I listen to the hymns which go back to the best memories of my childhood, then God is to me an emotional and comforting God—a protecting father.

When I am trying to work out a problem of right and wrong, then God is a clear and unambiguous voice, an unfailing source of moral guidance. I do not in the least understand how these things happen; but I know perfectly well, if I listen to this voice, what is the right thing to do. I have many times been uncertain which course of action would best serve a certain *practical* purpose: but I cannot think of a single instance in my life when I asked what was the really *right* thing to do and the answer was not forthcoming.

These two statements cover my everyday relation with God. I do not find it helpful—or necessary—to try to analyze these statements in logical terms. They state facts of *experience*. You can no more convince me that there is no such God than you can convince me that a table or a rock is not solid—in each case the evidence is simple, direct, and uniform.

As a scientist who is familiar with the detailed explanations of the atomic structure of, say, the table and the rock, it does not surprise me, or disturb me, that these everyday concepts of God do not offer me detailed logical explanation. God on an intellectual plane (corresponding to the theoretical plane of the physicist) is something else. That "something else," just as a scientist would expect, is very abstract: On the intellectual level, God is, to me, the name behind a consistent set of phenomena, which are all recognizable in terms of moral purpose and which deal with the control of man's destiny. I shall explain this in greater detail in a moment.

Can a scientist believe in God? Some persons think that scientists simply can't believe in God. But I think scientists have unique advantages here, for scientists are precisely the persons who believe in the unseeable, the essentially undefinable.

No scientist has ever seen an electron. No scientist soberly thinks that anyone ever could. In fact, "electron" is simply the name for a consistent set of things that happen in certain circumstances. Yet nothing is more "real" to a scientist than an electron. Chairs and tables and rocks—these are, in fact, not very "real" to a scientist if he is thinking deeply. A table, viewed with the precise tools of the atomic physicist, is a shadowy, swirling set of electric charges, these electric charges themselves being vague and elusive. So viewed, the table completely loses its large-scale illusion of solidity. In fact, the modern scientist has two sets of ideas about the world, which he carries in his head simultaneously. He uses the simpler set of ideas when it works, and he falls back on the more fundamental set when necessary. The simpler set of ideas deals with large-scale objects—you, me, tables, chairs, rocks, mountains. For these large-scale objects the scientist has a workaday set of ideas about solidity, location, reality, etc. In these everyday terms, a rock is solid and real because it hurts your toe when you kick it. You know how to measure where a star is and how it is moving. These ideas are extremely useful. If a scientist got up some morning without these workaday ideas, he couldn't even succeed in getting his shoes on. Indeed, he would never figure out how to get out of bed.

But the scientist also knows that all these large-scale ideas simply *do not stand up under close examination.* When he forces his thinking down to basic levels, a wholly new and strangely abstract set of ideas comes into play. Solids are not really solid. "Real objects" are not even composed, as physicists thought a half century ago, of submicroscopic atoms like billiard balls.

Consider the electron, for example. For a while physicists thought it was a particle. (You mustn't really ask what "particle" means, any more than you should ask just what it means when you say God has certain human characteristics.) Then physicists realized that electrons are wave motions. (Wave motions of what? Well, it isn't useful to ask this question, either.) Today, physicists think of electrons as being both (or either) particles or waves.

Further, you can't pin down this electron object, whatever it is. If you ask the electron more and more insistently, "Where are you?" you end up with less and less information about where it *is*. I am not being facetious. Modern physics simply cannot tell both where a particle is and where it is going; it can answer one or the other, but not both.

Or suppose you carry out careful measurements and consult the best theories of physics to determine what an electron is going to do next. Well, it turns out that you can only say what it is *likely* to do next. Science can predict with great definiteness on large-scale, everyday sort of phenomenon; but this definiteness fades away and vanishes as you proceed down the scale of size to individual events. If a scientist is studying just two electrons, it turns out to be completely hopeless for him even to try to keep track of which is which.

All this may seem funny or ridiculous to you. But you had better not jump to unwise conclusions. Science may move on to more advanced views of the ultimate nature of things, but there is not the slightest promise that the "improved" view can be any less abstract. More scientists, I think, have had to come to an entirely new concept of what "explaining," "understanding," or "defining" really mean. And this holds for science no less than religion.

Let us take stock of where we are. I am trying to explain whether or not a scientist can believe in God. To do this, I am trying to explain the way scientists think. And we find that a scientist is, by his training, especially prepared to think about things in two ways: the commonplace, everyday way, and a second way which is a deep, logical, restless, and detailed way. In this second way of thinking the scientist is forced to live with very abstract ideas. He has come to feel their value and their inevitability. He has developed skepticism concerning easy answers or the "obvious" nature of events. He is the last to expect that an "ultimate explanation" is going to involve familiar ideas. He is convinced, moreover, that reality is not simply denseness or visibility, hardness or solidity. To the scientist, the real is simply what is *universally experienced*.

Does this sound abstract and difficult? Of course it does. The scientist knows that when he is pushed back to a point at which his thinking should begin, he is forced to deal with difficult abstractions. A scientist is just the one who should not say that an abstract concept of God results in an "unreal" God. For the scientist knows that the everyday reality of the table and the rock is an illusion, and that reality is in fact a very subtle, evasive, and somewhat abstract business.

A scientist does not accept ideas just because they are abstract or unreal. He raises a very basic question: "Does this definition *work* successfully?" "Electron" is only the name behind a set of phenomena, but essentially all physicists agree as to what these electron phenomena are; and there is a high degree of agreement on the rules which govern electron phenomena. If there is this kind of consistency, then a definition "works"—and the scientist finds it acceptable and satisfying. Man has not attained the same universal agreement, or consistent explanations, for what can be called God phenomena. Yet I accept the idea of God for three reasons:

First, in the total history of man there has been a most impressive amount of general agreement about the existence (if not the details) of "God." This agreement is not so logically precise as the agreements about electrons; but far, far more people believe and have believed in God than believe or have ever believed in electrons.

Second, I know I cannot think through the realm of religious experience as satisfactorily as I can think through certain smaller and less important problems. But the nuclear physicist himself has only incomplete and contradictory theories. The theories work pretty well and represent the best knowledge we have on a very important subject.

Third, I accept two sets of ideas of God—the everyday concept of an emotional and intuitive God, and the intellectual concept of an abstract God—for the very solid reason that I find both of them personally satisfying. It does not at all worry me that these are two rather different sets of ideas; if an electron can be two wholly inconsistent things, it is a little narrow to expect so much less of God.

Can a scientist believe the Bible? I think that God has revealed Himself to many at many times and in many places. I think, indeed, that He keeps continuously revealing Himself to man today. *Every new discovery of science is a further "revelation" of the order which God has built into His universe.*

I believe that the Bible is the purest revelation we have of the nature and goodness of God. It seems to me natural, indeed inevitable, that the human record of divine truth should exhibit a little human frailty along with much divine truth. It seems to me quite unnecessary to be disturbed over minor eccentricities in the record.

There are, of course, sincere and earnest persons who seem to find it necessary to place a literal interpretation on every word in the Bible, and who accept every statement as divinely revealed truth. This attitude seems to me to lead to both spiritual and intellectual poverty. The reports of miraculous happenings in biblical times seem to me more reasonably understandable as poetic exaggeration, as ancient interpretations of events which we would not consider miraculous today, or as concessions (on the part of Christian writers) to the problem of competing with the magical claims of other religions.

Can a scientist believe in miracles? Put a kettle of water on the stove. What happens? Does the water get hot and boil, or does it freeze? The nineteenth-century scientist would have considered it ridiculous to ask this question. But scientists today, aware of the peculiarities of modern physical theories, would say: "In the overwhelming proportion of the cases, the water will get hot and boil. But in one of a vast number of trials, it is to be expected that the water will *freeze* rather than boil."

Modern science recognizes the exceedingly rare possibility of happenings—such as water freezing on a hot stove or like a brick spontaneously moving upward several feet—which so contradict the usual order of events that they can be called "miracles." No one can logically hold that science rules out "miracles" as impossible.

If my religious faith required miracles, my scientific knowledge would not necessarily deny them. But my religious faith does not at all rest on the validity of ancient miracles. To me, God gains in dignity and power through manifestations of His reason and order, not through exhibitions of caprice.

Can a scientist believe in "life after death"? Scientists are very heavily (but not exclusively, as some assert) influenced by evidence: If there is good evidence for a statement, they accept or believe the statement; if there is good evidence against, they reject. If it seems impossible to produce any evidence—either for or against a statement—then scientists tend to consider such statements as unprofitable matters of inquiry.

So far as I am concerned, "life after death" is a matter in which I can neither believe nor disbelieve. Until now, at least, I have been too much interested in this life on earth to feel any urge to indulge in pure speculation about another.

4 General Theory of Relativity and Buddhist Views of Time and Space

Speak of time and people think of the movement of a clock. Philosophers make ideological speculations on time and space, and astronomers and other scientists remember the relative movements of heavenly bodies, the velocity of light and the general theory of relativity.

From ancient times, the problem of time and space was considered extremely difficult. It is yet to be solved.

In Buddhism, however, the deep theory of time was advocated almost 3,000 years ago. It is not too much to say that time as explained in Buddhism is the true theory of time. Some of the Buddhist theories on time are: the time when a Buddha responds to the nature of the people; the three periods of *Shoho*, *Zoho* and *Mappo*;* and the time as is shown in the principle of Five Times

---

* *Shoho, Zoho* and *Mappo:* The period of 1,000 years after Sakyamuni's death is known as *Shoho* and the following 1,000 years known as *Zoho*. The period 2,000 years after Sakyamuni's death when his Buddhism has lost the power of salvation and when the True Buddha, Nichiren Daishonin, has made His advent to save all mankind from unhappiness.

192

and Eight Teachings.* Besides, there is the Buddhist principle that life has neither beginning nor end, or "Fifty *Sho Ko* are as if they were half a day,"* as written in the *Yujuppon*, Chapter 15 of *Hokekyo*. However, Nichiren Daishonin's theories on time, among others, are the highest level attainable by philosophy. They include *Kuon Ganjo*,* *Kuon soku Mappo*,* *Ku-Matsu Ichido*,* etc. Einstein's theory of time, it seems to me, is a part of the Buddhist theory.

Buddhism explains time profoundly from the viewpoint of human activities—the function of life. That is to say, Buddhism makes much of the time we feel in our everyday life. This is a start in our pursuit of the essence of time. Time, according to Buddhism, is a specific character of life and is felt in the activities and changes of the universe. The time set by physics is to be used with the Buddhist ideas as its basis.

In this section, I will discuss, first, the physical time we use in our daily life, and secondly, Newton's theory of absolute time, as well as Einstein's theory of relative time, and thirdly, the essence of time as made clear in Buddhism.

First of all, what is the time we use in our daily life? Time has

---

* Five Times and Eight Teachings: Sakyamuni's preaching is classified into five periods—*Kegon, Agon, Hoto, Han'nya,* and *Hokke-Nehan*. His teaching is divided into two—its contents and its form of preaching. Each of them is subdivided into four. Hence 'Eight Teachings'.

* Fifty *Sho Ko* ...: A passage from *Yujuppon* (the fifteenth chapter about the issuing of Bodhisattvas from the earth) reads: "And while those Bodhisattvas Mahasattvas who had emerged from the gaps of the earth were saluting and celebrating the *Tathagatas* by various Bodhisattva hymns, fifty intermediate *kalpas* in full rolled away, during which fifty immediate kalpas the Lord Sakyamuni remained silent, and likewise the four classes of the audience. Then the Lord produced such an effect of magical power that the four classes fancied that it had been no more than one afternoon, . . ." 'Kalpas' (*Ko*) is, according to a theory, eight million years. It indicates that the four classes including Bodhisattvas of pre-Hokekyo teachings were not fully enlightened, while Bodhisattvas from the earth obviously could see the fact.

* *Kuon Ganjo:* It is indicative of the time when Nichiren Daishonin attained enlightenment, indescribably long before the time (called simply 'Kuon') when Sakyamuni Buddha was enlightened in his past existence. *Kuon Ganjo* in the life-philosophy has a deeper meaning. (See *Kuon Jitsujo* in page 200.)

* *Kuon soku Mappo, Kumatsu Ichido*: 'Kuon' (Aeon) or 'Ku' of 'Kumatsu' means *Kuon Ganjo* when people believed in Nam-myoho-renge-kyo. In 'Mappo' or 'Matsu' of 'Kumatsu,' Nam-myoho-renge-kyo also spreads. Therefore, the two periods are the same ('soku' or 'Ichido') in the teaching of salvation.

two meanings. It first indicates a specific moment like "I came to the office at 8:30 a.m.," and secondly the continuance of time, such as "the length of time for an examination is three hours." Such time is defined by the revolution of the earth around the sun and its rotation around its axis.

We do not find it inconvenient to live according to Newton's absolute time. Such time is fixed on the presumption that time passes at a constant rate and that the earth continues its revolution and rotation with the same speed. It takes a year for the earth to move around the sun, bringing forth the four seasons of spring, summer, fall, and winter. A day is the length of time when the rotation of the earth brings about a day and a night.

Ancient Egyptians divided the day into 24 hours. They lived with the movement of the sun as their standard of living. Around 4,000 B.C., they used the solar calendar having 365 days in a year. About 3,000 B.C., they divided the day and the night into 12 hours each.

In Babylonia, they also set time by the same method. Another method divided the day and the night into 12 hours. These ideas were developed in ancient Greece and Rome. In China and Japan, twelve zodiacal signs* were used as divisions of time.

The time measurement we now use divides a day into 24 hours, an hour into 60 minutes, and a minute into 60 seconds. We use the mean solar day* and the mean solar time instead of the true solar day and the apparent solar time. We do not use the true solar day which is measured by the apparent daily motion of the sun from east to west. The interval between the two successive transits of the apparent sun across the meridian* is called an apparent solar

---

* Twelve zodiacal signs: The designation of time in ancient Japan. The day and night are divided into twelve periods and each is called by the name of twelve animals—mouse (midnight), ox, tiger, hare, dragon, snake, horse (noon), sheep, monkey, cock, dog and wild boar.
* Mean Solar Day (Time): The 'mean solar' is a fictitious sun which moves at a uniform rate in the celestial equator and has its right ascension always equal to the sun's mean longitude. The time recorded by the mean sun is mean solar or clock time; it is regular as distinct from the nonuniform solar sundial time. The adjective 'mean' is chiefly used in the sense of 'average' as in mean temperature, mean birth or death rate, etc.
* Meridian: See next page.

194

day. The earth moves about the sun in an elliptical orbit which is inclined to the equator. In consequence of these factors, the component of the daily motion of the sun along the equator varies. As a result, apparent solar days are of variable length, and apparent solar time is not uniform. The mean solar day is fixed with a fictitious sun moving along the ecliptic* at such a velocity that the length of the day is the same all the year round. The mean solar day is divided into 24 hours which are called the mean solar time. However, in daily use, the mean solar time is calculated from the mean sidereal time which is derived from the true sidereal time. A year is 365.2422 mean solar days or 365 days, 5 hours, 46 minutes, and 48 seconds.

Because of variations in the speed of the rotation of the earth, the second of the mean solar time is not a constant unit of time. Hence, a redefinition of the second as 1/31,556,925.9747 of the tropical year* for January 1900.

To measure with higher precision, the study of atomic clocks is being promoted, using as its mechanism the vibration of nitrogen atoms in ammonia molecules and the periodical vibration of cesium atoms.

In any case, there can be no absolutely correct time measured

---

* Meridian: Geographical meridians are the great circles drawn on the earth's surface which pass through the poles, and thus pass through all places having the same longitude. The meridian of Greenwich (the great circle passing through Greenwich Observatory and the two poles) is the zero from which terrestrial longitudes are reckoned. In astronomy, the meridian is the great circle through the pole and the zenith; it intersects the horizon in the north and south points.

* Ecliptic: Ecliptic, in astronomy, is the great circle on the celestial sphere which forms the apparent path of the sun in the course of the year. The twelve constellations or signs of the zodiac are arranged along the ecliptic. The plane of the ecliptic is the plane of the earth's orbit, or more strictly the plane in which the combined center of gravity of the earth and moon revolves around the sun; it meets the celestial sphere in the great circle mentioned above.

* Tropical year: Until 1955 the fundamental unit of time was the second of mean solar time, defined as 1/86,400 of the mean solar day. Because of variations in the speed of rotation of the earth the second of mean solar time is not a constant unit of time. A redefinition of the second as 1/31,556,925.9747 of the tropical year for 1900 January 1 at 12 hrs. E. T. was adopted by the International Committee on Weights and Measures in 1956, at Paris.

195

objectively, whether time is measured by the rotation and the revolution of the earth or by the vibrations of an atom.

In an isolated community, clocks and watches would be set to the local mean time. The local standard time of the world is defined by using Greenwich time at its standard. Greenwich time is set with twelve hours before the mean sun passes the Greenwich meridian as 0:00. Every country has its own standard time with differences determined by integral numbers. In Japan, the 135th degree of east longitude is set as the standard time. Japan Standard Time is nine hours faster than the Greenwich time.

The development of jets has made travelling abroad easy, but travel by jets often embarrasses us because of differences in time. Differences in time are as much as 10 or 15 hours between the local standard times. Twenty-four hours are a unit of life but in travelling overseas, the differences in time will disturb our regular life. To make oneself accustomed to a new life is rather difficult.

Secondly, what is the theory of relativity? Classical physics from Newton's time regarded time and space as absolute. However, Einstein's theory of relativity has replaced the old idea with the theory that time is relative.

Newton's classical dynamics was widely accepted until Einstein made public the theory of relativity. Newton advocated the absoluteness of time and space in his work, '*Principia*'.

However, Einstein asserted that time and space do not separate one from the other but are closely related to one another. Neither time nor space can exist independently. All things in the universe have time and space and continue to move and change. We live in a world of four dimensions with time being the fourth dimension. The fourth dimension is thus named because space has three dimensions of its own: length, width and depth.

In conclusion, time is relative and it cannot be measured in the same way at any place. The theory of relativity is deeply connected with light. The velocity of light is about 300,000 kilometers per second. Stars in the universe take tens of thousands of light years or hundreds of millions of light years to reach the earth. A star which is 100,000 light years away from the earth emitted the light which we can now see, 100,000 years ago. However, we do not feel the theory of relativity functioning in our daily life as the velocity of light is incomparably greater than any speed we have experienced. Of course, in actuality it is said that under the influence of the velocity of light we can find some strange phenomena on earth.

Suppose two passengers sitting in different seats of a train, light their cigarettes at the same moment. To a person standing on the ground it will appear as though the passengers had lit the cigarettes at two different moments. This will be understood if you imagine that it had happened on two different stars. Thus, two actions which take place in different spots at the same time will not seem to be simultaneous, but rather at different times, when they are observed from a place apart.

We cannot make a speed greater than that of light. As a body increases its speed, the passing of time becomes very slow. If it travels at the velocity of light, we cannot recognize the changes and movements of life and consequently we cannot perceive the passing of time.

According to G. Gamow, a Soviet theoretical physicist, if in the future a space ship travels from the earth to a planet of the solar system or a planet of a fixed star, its pilots and passengers, when they return, may be younger than the people of the same age who have stayed on earth. This can be proven mathematically if we take into consideration the acceleration resulting from the turn of the space craft.

Although it is unrealistic, if we were to travel in a space craft with 98 percent of the velocity of light to Sirius, which is eight light years away, we would be able to return to the earth in nine years, but calculation will show that the people on earth would feel the return of the space craft 16 years after its departure. Travelling to the center of the Galactic System (Milky Way) with a specific acceleration and returning to the earth will take 40,000 years, according to the calendar on earth, but the clock set to the space craft will show it to take only 30 years.

These are of course fictitious stories which show the relativity

197

of space and time, because the use of an ion rocket* or atomic energy cannot make possible travelling to a distant planet or a fixed star. Anyway the relativity of time and space is undeniable.

Indeed, such a phenomenon is observed in actuality. The elementary particles moving near the velocity of light, for example, are observed to exist 50 times as long as their span of life in a static condition.

Concretely, the μ neutron, one of the secondary cosmic rays*, exists only for one-millionth second and then changes into an electron and two neutrinos. The length of its life is constant and free from any outer influence. μ neutrons are made of π neutrons in the stratosphere some 15 kilometers above the earth and reach the earth at a speed 0.9998 times as fast as light (which travels at 300,000 kilometers a second). This is an obvious fact based on observation.

Therefore, μ neutrons must have extended their life 50 times longer than in its static condition, because they travel near the velocity of light. The maximum distance the π nuetrons can travel is one-millionth of 300,000 kilometers or 0.3 kilometers, but they fly 15 kilometers to reach the earth or 50 times of their maximum distance. This wonderful happening can be clarified by the theory of relativity. The rocket which travels near the speed of light shows the same result.

Thus, the relativity of time and space has and will exert great

---

* Ion rocket: By inducing ions to an electric field caused by electric power (which is generated by atomic energy), great energy can be produced for rockets. Ions in use are those of alkali metals such as cesium and calcium. Those ions work as accelerating medium.

* Secondary cosmic rays: The terms 'cosmic rays' refer both to primary and to secondary cosmic rays. Primary cosmic rays are submicroscopic, electrically charged particles—largely protrons—that travel in space at speeds nearly equal to that of light. Some of them happen to approach the earth where, high in the atmosphere, they collide with atoms in the air, giving part of their energy to electromagnetic radiation (photons) and to other particles, which proceed in nearly the same direction as the primaries. These new particles and photons are called secondary cosmic rays. Like the primaries, they too many interact with atoms in the air, or eventually atoms in the earth until ultimately the energy is converted into heat. The primary and secondary rays are not alwasy distinguishable from each other.

198

influence on other fields of learning as the space age develops. The time-space problem of the theory of relativity has also affected philosophies and thought. However, a study of Buddhist philosophy shows that the essence of Buddhism has revealed the theory of relativity or rather a more perfect theory of time-space.

We realize time through changes and movements of life in the universe. We decide time by using the almost regular movements of the earth—its rotation and revolution. The theory of relativity proved the righteousness of the Buddhist idea of time-space. After all, the theory is based mainly on the existence, changes and movements of life in the universe. Therefore, it is natural that the hypothesis of space which reverses history with a fictitious velocity that surpasses light can never be justified. From the viewpoint of Buddhist philosophy, phenomena which are knitted together by the law of cause and effect cannot be reversed. There can be no reversibility for them. On planets, which have the different speeds of rotation and revolution, the measurement of time is different from . that of the earth. Buddhist philosophy reveals a leading thought of contemporary science by explaining time as subjective with vital activities as its basis.

Buddhism asserts that time is felt by living things. When we are having a good time, time flies. We feel three and four hours pass by as if they were as many minutes. The Buddhist view of time is easily understood in our daily life.

On the contrary, when we are in agony, we feel the day is very long and wish that it would pass as quickly as possible. Under such circumstances, we feel as if time were dragging on. In a state of hell, we feel time passing so slowly that our agony increases.

Time, as realized through the activities of life, is different from time measured by a clock and its length is determined by the condition of a person. When we sleep a sound sleep, we are unconscious of time. Life after death in the universe will also be unconscious of time. Awaking next morning, we realize the past by remembering by chance some past events.

199

Ongi Kuden, a record of Nichiren Daishonin's oral teachings, reads:

"Third: On *Gajitsu Jobutsu Irai Muryo Muhen* (literally, "Actually, uncountable and unfathomable length of time has passed since I attained enlightenment.")—*Ongi Kuden* reads, '*Gajitsu*' means Buddha's attaining enlightenment in *Kuon*, the infinite past. However, the true meaning is that '*Ga*' is indicative of all living things in the universe or each of the Ten Worlds (*Jikkai*) and that '*Jitsu*' is defined as Buddha of *Musa Sanjin** . . . The person who realizes this is named Buddha. '*I*' (literally, already) means the past and '*rai*' (literally, to come) the future. '*Irai*' includes the present in it. Buddha has attained the enlightenment of '*Gajitsu*'* and His past and future are of uncountable and unfathomable length."

The same *Ongi Kuden* goes: "*Kuon* means having neither beginning nor end, being just as man is, and being natural. It has neither beginning nor end because *Musa Sanjin* is not created in its original form. It is just as man is because it is not adorned by the 32 wonderful physical features and the 80 favorable characters. It is natural because the Buddha of *Hon'nu Joju** is natural. *Kuon* is Nam-myoho-renge-kyo. *Kuon Jitsujo**—really enlightened. Enlightened as *Musa*."

In the quotation, the Daishonin has spoken about time—past, present and future. He has said, " '*I*' (literally, already) means the past and '*rai*' (literally, to come) the future. '*Irai*' includes the

---

* *Musa Sanjin*: *Musa* means not artificial but natural. Life comprises *Hosshin* (the essence of life), *Hoshin* (wisdom) and *Ojin* (body). These three phases of life continue to exist eternally. *Musa Sanjin* is another name for True Buddha.

* *Gajitsu*: Literally, 'Ga' means I or ego, and 'jitsu' True Buddha (called *Musa Sanjin*). The quotation "Buddha has attained the enlightenment of 'Gajitsu' and . . ." means that Buddha has realized that he himself is *Musa Sanjin* (Buddha). 'Jitsu' (actually or true) means that Hokekyo, (now, Gohonzon) is the true teaching.

* *Hon'nu Joju*: Life is eternal and has neither beginning nor end.

* *Kuon Jitsujo*: Literally, Buddha attained enlightenment in the Kuon, infinite *Aeon*. Nichiren Daishonin clarified its true meaning as "enlightened as *Musa*" which means that the True Buddha in Mappo has made His advent as a common mortal or 'bompu' which is *Musa* as distinguished from *Usa* (adorned by various idealistic features).

present in it. Buddha has attained the enlightenment of '*Gajitsu*' and His past and future are of uncountable and unfathomable length." This is the important principle that a present moment includes both the past and future. *Sokanmon Sho*,* Nichiren Daishonin's writing, reads, "The three times of past, present and future are the reasons of a momentary existence of life; therefore, they are inseparable." Time can be classified into three, past, present and future expediently, but there is no distinction among them from a viewpoint of the essence of life.

The ultimate movement of life is in the present moment. The past can be known by recalling to oneself of past events. Without remembrance there cannot be the past. To remember something in the past is to find in a present moment the vital activities of the past. Therefore, we have the past because we have this moment. The past is included in the present moment. Also to think of what will happen in the future, recognize and believe in the future is the function of life in the present moment. The moment we regard as the present changes into the past in the next moment. A moment in the future soon becomes present and passes into the past. In conclusion, the past, present and future exist in this moment of life.

I. Kant, a great German philosopher, has said that time is a subjective matter existing in our minds.

Kant's predecessors speculated that time and space were objective existences. By unifying all their philosophies, Kant established a new philosophical idea. Kant said that time and space do not exist objectively in the outer world but that they are subjective forms for observing things in the outer world. He boasted that he had reversed the conventional philosophy and that his achievement should be likened to that of Copernicus who founded the heliocentric theory. Kant's time-space theory was a step forward toward Buddhism.

---

* *Sokanmon Sho*:   The full name of this writing is 'Sanze-no Shobutsu Sokanmon Sho.' *Sanze-no Shobutsu* (all the Buddhas of the three existences of life, past, present and future) proved unanimously that Nam-myoho-renge-kyo is the only true teaching. Sokanmon Sho also theoretically clarifies the real aspect of life.

A French philosopher, H. Bergson, (1859–1941), has said: "Consciousness is a connection between what there was and what there will be, or a bridge between the past and the future." He has also said, "Present in its strict sense is an indivisible moment—the moment which is considered infinitely small, always moving, at once coming into existence and coming to an end." This excellent view of time is a part of Buddhism.

The late president Toda has lectured: "The moment we think the present changes into the past and the moment we consider still in the future immediately turns into the present and then into the past without any hesitation. If we want to say there is a moment, it has passed already and that particular moment is no longer. If we deny the existence of a moment, there is a moment now. It is equal to the idea of 'Ku'. We feel happy or unhappy, become hopeful or disappointed in such a moment. This moment can be called the entity of life."

Furthermore, a moment of life has the law of causality. All causes in the past become present effects and present causes become future effects. A Mystic Law which carries cause and effect simultaneously is called *Nam-myoho-renge-kyo*.*

Ongi Kuden reads: "There cannot be Six Paths and Four Saints* anywhere but in the momentary existence of life which is the ultimate essence of life or in any one of the Ten Worlds. *Nam-myoho-renge-kyo* is, after all, but a moment in which three existences of life, the past, present and future are included." *Hyaku Rokka Sho*, Nichiren Daishonin's writing on one hundred and six arcana of Bud-

---

* *Nam-myoho-renge-kyo:* The 'renge' of this invocation means a lotus flower which carries a flower (cause) and a pulp (effect) at the same time.

* Six Paths and Four Saints: It is significant of *Jikkai* (Ten Worlds). The former six worlds are known as Six Paths, and the latter four as Four Saints. The former include *Jigoku* (Hell), *Gaki* (Hunger), *Chikusho* (Animality), *Shura* (Anger), *Nin* (Tranquility) and *Ten* (Rapture). These six states we can experience every day. The Four Saints include *Shomon* (Learning), *Engaku* (Absorption), *Bosatsu* (Bodhisattva) and *Butsu* (Buddha). The Ten Worlds appear in a momentary state of life one by one. There can be none of the Ten Worlds without a moment. The moment is the ultimate existence.

* *Hyaku Rokka Sho:* Literally, a writing on one hundred and six items. Nichiren Daishonin wrote the essential teachings of the True Buddhism as are distinguished from Sakyamuni's Buddhism for Nikko Shonin, the successor to the Daishonin. The writing has been handed down in Nichiren Shoshu religion.

202

dhism, goes; "*Kuon* or eternity is included in a momentary existence of life, the essence of Buddhism. Such is a Mystic Law . . . " *Hon'nin Myo Sho*,* the Daishonin's writing on the depths of His Buddhism, reads: "*Kuon* or eternity in a momentary existence of life, is *Nam-myoho-renge-kyo*."

Thus, both past and future are included in this moment. Eternity is a succession of vital activities changing from moment to moment. This moment includes life of the endless past and life of eternity. The life of *Kuon* exists in a momentary existence of life.

Kuon soku Mappo—"*Kuon* means having neither beginning nor end, being just as man is, and being natural." The essence of vital activities is, after all, a moment. This is termed either *Kuon Ganjo* or *Mappo*. Both *Kuon* and *Mappo* are the periods when *Nam-myoho-renge-kyo* of *San-dai-hiho*,* Three Great Secret Laws, spreads. Thus, *Kuon* is equal to *Mappo*. It is also known as *Ku-Matsu Ichido*, or, *Kuon* and *Mappo* are the same.

From a Buddhist viewpoint, there is no time without our vital activities. Time in its true sense is for life to perceive the movements and changes of life in the universe. Time as revealed in Buddhism is the true idea of time.

---

* *Hon'nin Myo Sho*: Nichiren Daishonin interpreted in this writing the secret teachings of Dengyo the Great of Japan from a viewpoint of the True Buddhism. Thus He clearly distinguished His Buddhism from Sakyamuni's. Dengyo's teachings are still within Sakyamuni's Buddhism.
* *San-dai-hiho*, Three Great Secret Laws: They are *Honmon-no-Honzon* (the object of worship), *Honmon-no Daimoku* (the invocation of Nam-myoho-renge-kyo) and *Honmon-no Kaidan* (the sanctuary). Honmon means the True Buddhism and 'no' means 'of.' These three laws are indispensable to the practice of the True Buddhism.

Chapter Three

THE INFLUENCE OF LANGUAGE

Culture and language have such a strong and all-embracing hold on each one of us that it is difficult to identify their role in shaping our thoughts and points of view. Only by comparing our ideas with those of people who do not share our culture and/or language can we become aware of fundamental beliefs we never question. Sometimes one gets the feeling that modern science is a product of Western Europe which people of different traditions and background can assimilate only with difficulty.

Benjamin Lee Whorf's extensive studies of American Indian languages led him to speculate about the relation of words and phrases to concepts, theories, and a picture of the world. He concludes that the mere use of language introduces a bias into descriptions of nature and constrains the interpretations scientists can make. Whorf's ideas will become more real for you if you can discuss them with colleagues whose linguistic background is very different from yours.

The other two articles in this chapter are devoted to children's thinking and reflect cultural factors; they do not refer directly to language. The discussion of the level of water, which we have selected from Piaget's voluminous writings, concerns a particularly simple phenomenon. You will have no difficulty duplicating the investigation with children in your community. Yet the example clearly shows the development in children's thinking and the deviations from adult or scientific explanations you will encounter. Surely educators might be expected to recognize these differences in their planning of curricula. Do they? Your experience in science courses should give you a partial answer to this question.

The problems which faced Francis Dart and Panna Lal Pradhan when they attempted to influence science education in Nepal led them to investigate children's thinking about natural phenomena. The two men found an undigested mixture of myth and school learning on which it was difficult to build further learning such as scientists or technicians might require. As a matter of fact, Piaget has obtained rather similar responses from young children in Geneva but found these to disappear when the children become older. In a way, therefore, the confrontation of science with culture may not be resolved as neatly and intellectually as Weaver and Ikeda reconciled science and religion (see Chapter 2).

About the Contributors

BENJAMIN LEE WHORF (1897-1941) was fire prevention inspector and an executive with the Hartford Fire Insurance Company. He was educated as a chemical engineer, but an early interest in religion and the apparent discrepancy between the biblical and scientific accounts of the origin of the universe led him to an intense study of languages. Whorf thereupon embarked on a highly productive career in linguistic scholarship concurrent with his insurance work. He is especially noted for his investigations of American Indian languages and the general thesis that culture and language interact intimately.

JEAN PIAGET (1896-) is Co-Director of the Institut Jean Jacques Rosseau at the University of Geneva. Though educated in biology, he very early became interested in reasoning and has devoted his long and extremely productive career to the study of intelligence. He originated the concept of genetic epistemology, according to which a child's ability to reason passes through stages, with each stage forming a foundation for the next.

FRANCIS E. DART (1914-) is Professor of Physics at the University of Oregon, Eugene. He was born in Southern Rhodesia and studied physics in the United States. His principal interests have been the teaching of physical science and research in the physics of solids. Professor Dart has traveled widely in Asia on behalf of public and private agencies, and he developed a special concern regarding the interaction of science with non-European cultures.

PANNA LAL PRADHAN (1933-) is Reader, Head of the Department of Education, and Dean of the Faculty of Arts at Tribhuvan University, Kathmandu, Nepal. He is a native Nepalese but has studied extensively in India and the United States. His special research interests are in quantitative psychology and perception.

130

Science and Linguistics

E very normal person in the world, past infancy in years, can and does talk. By virtue of that fact, every person—civilized or uncivilized—carries through life certain naïve but deeply rooted ideas about talking and its relation to thinking. Because of their firm connection with speech habits that have become unconscious and automatic, these notions tend to be rather intolerant of opposition. They are by no means entirely personal and haphazard; their basis is definitely systematic, so that we are justified in calling them a system of natural logic—a term that seems to me preferable to the term common sense, often used for the same thing.

According to natural logic, the fact that every person has talked fluently since infancy makes every man his own authority on the process by which he formulates and communicates. He has merely to consult a common substratum of logic or reason which he and everyone else are supposed to possess. Natural logic says that talking is merely an incidental process concerned strictly with communication, not with formulation of ideas. Talking, or the use of language, is supposed only to "express" what is essentially already formulated nonlinguistically. Formulation is an independent process, called thought or thinking, and is supposed to be largely indifferent to the nature of particular languages. Languages have grammars, which are assumed to be merely norms of conventional and social correctness, but the use of language is supposed

* Reprinted from *Technol. Rev.*, 42:229–231, 247–248, no. 6 (April 1940).

to be guided not so much by them as by correct, rational, or intelligent THINKING.

Thought, in this view, does not depend on grammar but on laws of logic or reason which are supposed to be the same for all observers of the universe—to represent a rationale in the universe that can be "found" independently by all intelligent observers, whether they speak Chinese

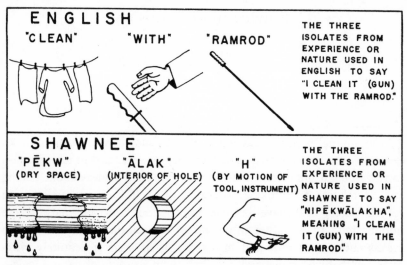

Figure 9. Languages dissect nature differently. The different isolates of meaning (thoughts) used by English and Shawnee in reporting the same experience, that of cleaning a gun by running the ramrod through it. The pronouns 'I' and 'it' are not shown by symbols, as they have the same meaning in each language. In Shawnee ni- equals 'I'; -a equals 'it.'

or Choctaw. In our own culture, the formulations of mathematics and of formal logic have acquired the reputation of dealing with this order of things: i.e., with the realm and laws of pure thought. Natural logic holds that different languages are essentially parallel methods for expressing this one-and-the-same rationale of thought and, hence, differ really in but minor ways which may seem important only because they are seen at close range. It holds that mathematics, symbolic logic, philosophy, and so on are systems contrasted with language which deal directly with this realm of thought, not that they are themselves specialized extensions of language. The attitude of natural logic is well shown in an old quip about a German grammarian who devoted his whole life to the study of the dative case. From the point of view of natural logic, the dative case and grammar in general are an extremely minor issue. A different attitude is said to have been held by the ancient Arabians: Two princes, so the story goes, quarreled over the honor of putting on the shoes of the most learned grammarian of the realm; whereupon their father, the caliph, is said to have remarked that it was the glory of his kingdom that great grammarians were honored even above kings.

The familiar saying that the exception proves the rule contains a good deal of wisdom, though from the standpoint of formal logic it became an absurdity as soon as "prove" no longer meant "put on trial." The old saw began to be profound psychology from the time it ceased to have standing in logic. What it might well suggest to us today is that, if a rule has absolutely no exceptions, it is not recognized as a rule or as anything else; it is then part of the background of experience of which we tend to remain unconscious. Never having experienced anything in contrast to it, we cannot isolate it and formulate it as a rule until we so enlarge our experience and expand our base of reference that we encounter an interruption of its regularity. The situation is somewhat analogous to that of not missing the water till the well runs dry, or not realizing that we need air till we are choking.

For instance, if a race of people had the physiological defect of being able to see only the color blue, they would hardly be able to formulate the rule that they saw only blue. The term blue would convey no meaning to them, their language would lack color terms, and their words denoting their various sensations of blue would answer to, and translate, our words "light, dark, white, black," and so on, not our word "blue." In order to formulate the rule or norm of seeing only blue, they would need exceptional moments in which they saw other colors. The phenomenon of gravitation forms a rule without exceptions; needless to say, the untutored person is utterly unaware of any law of gravitation, for it would never enter his head to conceive of a universe in which bodies behaved otherwise than they do at the earth's surface. Like the color blue with our hypothetical race, the law of gravitation is a part of the untutored individual's background, not something he isolates from that background. The law could not be formulated until bodies that always fell were seen in terms of a wider astronomical world in which bodies moved in orbits or went this way and that.

Similarly, whenever we turn our heads, the image of the scene passes across our retinas exactly as it would if the scene turned around us. But this effect is background, and we do not recognize it; we do not see a room turn around us but are conscious only of having turned our heads in a stationary room. If we observe critically while turning the head or eyes quickly, we shall see, no motion it is true, yet a blurring of the

HOPI - ONE WORD (MASA'YTAKA)
ENGLISH - THREE WORDS

ENGLISH - ONE WORD (SNOW)
ESKIMO - THREE WORDS

HOPI - PĀHE
ENGLISH - ONE WORD (WATER); HOPI - TWO WORDS

HOPI - KĒYI

Figure 10. Languages classify items of experience differently. The class corresponding to one word and one thought in language A may be regarded by language B as two or more classes corresponding to two or more words and thoughts.

scene between two clear views. Normally we are quite unconscious of this continual blurring but seem to be looking about in an unblurred world. Whenever we walk past a tree or house, its image on the retina changes just as if the tree or house were turning on an axis; yet we do not see trees or houses turn as we travel about at ordinary speeds. Sometimes ill-fitting glasses will reveal queer movements in the scene as we look about, but normally we do not see the relative motion of the environment when we move; our psychic makeup is somehow adjusted to disregard whole realms of phenomena that are so all-pervasive as to be irrelevant to our daily lives and needs.

Natural logic contains two fallacies: First, it does not see that the phenomena of a language are to its own speakers largely of a background character and so are outside the critical consciousness and control of the speaker who is expounding natural logic. Hence, when anyone, as a natural logician, is talking about reason, logic, and the laws of correct thinking, he is apt to be simply marching in step with purely grammatical facts that have somewhat of a background character in his own language or family of languages but are by no means universal in all languages and in no sense a common substratum of reason. Second, natural logic confuses agreement about subject matter, attained through use of language, with knowledge of the linguistic process by which agreement is attained: i.e., with the province of the despised (and to its notion superfluous) grammarian. Two fluent speakers, of English let us say, quickly reach a point of assent about the subject matter of their speech; they agree about what their language refers to. One of them, A, can give directions that will be carried out by the other, B, to A's complete satisfaction. Because they thus understand each other so perfectly, A and B, as natural logicians, suppose they must of course know how it is all done. They think, e.g., that it is simply a matter of choosing words to express thoughts. If you ask A to explain how he got B's agreement so readily, he will simply repeat to you, with more or less elaboration or abbreviation, what he said to B. He has no notion of the process involved. The amazingly complex system of linguistic patterns and classifications, which A and B must have in common before they can adjust to each other at all, is all background to A and B.

These background phenomena are the province of the grammarian—or of the linguist, to give him his more modern name as a scientist. The word linguist in common, and especially newspaper, parlance means something entirely different, namely, a person who can quickly attain agreement about subject matter with different people speaking a number of different languages. Such a person is better termed a polyglot or a multilingual. Scientific linguists have long understood that ability to speak a language fluently does not necessarily confer a linguistic knowledge of it, i.e., understanding of its background phenomena and its systematic processes and structure, any more than ability to play a good game of billiards confers or requires any knowledge of the laws of mechanics that operate upon the billiard table.

The situation here is not unlike that in any other field of science. All real scientists have their eyes primarily on background phenomena that cut very little ice, as such, in our daily lives; and yet their studies have a way of bringing out a close relation between these unsuspected realms of fact and such decidedly foreground activities as transporting goods, preparing food, treating the sick, or growing potatoes, which in time may become very much modified, simply because of pure scientific investigation in no way concerned with these brute matters themselves.   Linguistics presents a quite similar case; the background phenomena with which it deals are involved in all our foreground activities of talking and of reaching agreement, in all reasoning and arguing of cases, in all law, arbitration, conciliation, contracts, treaties, public opinion, weighing of scientific theories, formulation of scientific results.   Whenever agreement or assent is arrived at in human affairs, and whether or not mathematics or other specialized symbolisms are made part of the procedure, THIS AGREEMENT IS REACHED BY LINGUISTIC PROCESSES, OR ELSE IT IS NOT REACHED.

As we have seen, an overt knowledge of the linguistic processes by which agreement is attained is not necessary to reaching some sort of agreement, but it is certainly no bar thereto; the more complicated and difficult the matter, the more such knowledge is a distinct aid, till the point may be reached—I suspect the modern world has about arrived at it—when the knowledge becomes not only an aid but a necessity. The situation may be likened to that of navigation.   Every boat that sails is in the lap of planetary forces; yet a boy can pilot his small craft around a harbor without benefit of geography, astronomy, mathematics, or international politics.   To the captain of an ocean liner, however, some knowledge of all these subjects is essential.

When linguists became able to examine critically and scientifically a large number of languages of widely different patterns, their base of reference was expanded; they experienced an interruption of phenomena hitherto held universal, and a whole new order of significances came into their ken.   It was found that the background linguistic system (in other words, the grammar) of each language is not merely a reproducing instrument for voicing ideas but rather is itself the shaper of ideas, the program and guide for the individual's mental activity, for his analysis of impressions, for his synthesis of his mental stock in trade.   Formulation of ideas is not an independent process, strictly rational in the old sense, but is part of a particular grammar, and differs, from slightly to greatly, between different grammars. We dissect nature along lines laid down by our native languages. The categories and types that we isolate from the world of phenomena we do not find there because they stare every observer in the face; on the contrary, the world is presented in a

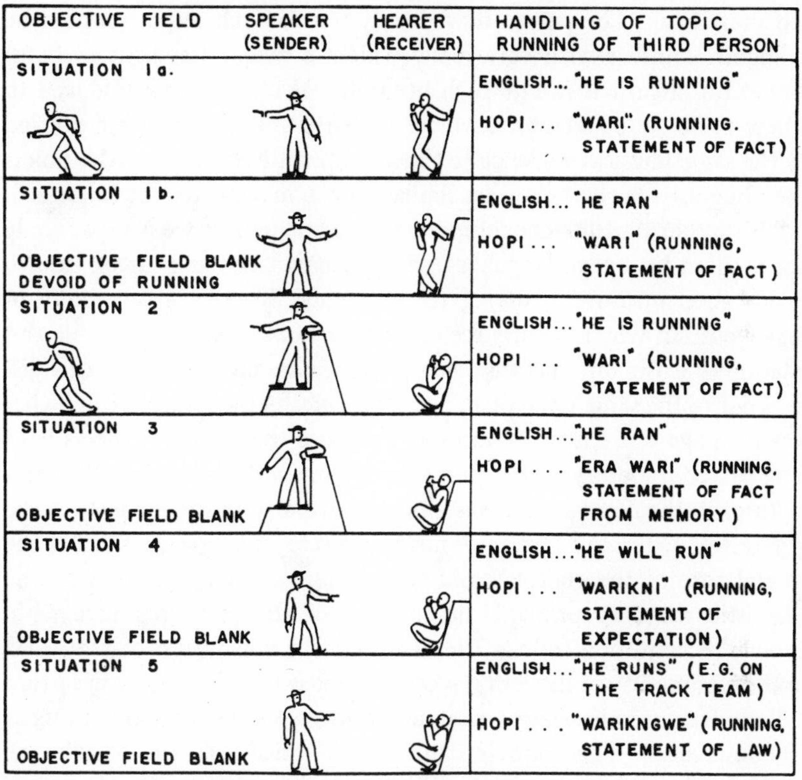

OBJECTIVE FIELD	SPEAKER (SENDER)	HEARER (RECEIVER)	HANDLING OF TOPIC, RUNNING OF THIRD PERSON
SITUATION I a.			ENGLISH... "HE IS RUNNING" HOPI... "WARI." (RUNNING. STATEMENT OF FACT)
SITUATION I b. OBJECTIVE FIELD BLANK DEVOID OF RUNNING			ENGLISH... "HE RAN" HOPI... "WARI" (RUNNING, STATEMENT OF FACT)
SITUATION 2			ENGLISH..."HE IS RUNNING" HOPI.... "WARI" (RUNNING, STATEMENT OF FACT)
SITUATION 3 OBJECTIVE FIELD BLANK			ENGLISH..."HE RAN" HOPI... "ERA WARI" (RUNNING. STATEMENT OF FACT FROM MEMORY)
SITUATION 4 OBJECTIVE FIELD BLANK			ENGLISH..."HE WILL RUN" HOPI... "WARIKNI" (RUNNING, STATEMENT OF EXPECTATION)
SITUATION 5 OBJECTIVE FIELD BLANK			ENGLISH..."HE RUNS" (E.G. ON THE TRACK TEAM) HOPI.... "WARIKNGWE" (RUNNING, STATEMENT OF LAW)

Figure 11. Contrast between a "temporal" language (English) and a "timeless" language (Hopi). What are to English differences of time are to Hopi differences in the kind of validity.

kaleidoscopic flux of impressions which has to be organized by our minds—and this means largely by the linguistic systems in our minds. We cut nature up, organize it into concepts, and ascribe significances as we do, largely because we are parties to an agreement to organize it in this way—an agreement that holds throughout our speech community and is codified in the patterns of our language. The agreement is, of course, an implicit and unstated one, BUT ITS TERMS ARE ABSOLUTELY

OBLIGATORY; we cannot talk at all except by subscribing to the organization and classification of data which the agreement decrees.

This fact is very significant for modern science, for it means that no individual is free to describe nature with absolute impartiality but is constrained to certain modes of interpretation even while he thinks himself most free. The person most nearly free in such respects would be a linguist familiar with very many widely different linguistic systems. As yet no linguist is in any such position. We are thus introduced to a new principle of relativity, which holds that all observers are not led by the same physical evidence to the same picture of the universe, unless their linguistic backgrounds are similar, or can in some way be calibrated.

This rather startling conclusion is not so apparent if we compare only our modern European languages, with perhaps Latin and Greek thrown in for good measure. Among these tongues there is a unanimity of major pattern which at first seems to bear out natural logic. But this unanimity exists only because these tongues are all Indo-European dialects cut to the same basic plan, being historically transmitted from what was long ago one speech community; because the modern dialects have long shared in building up a common culture; and because much of this culture, on the more intellectual side, is derived from the linguistic backgrounds of Latin and Greek. Thus this group of languages satisfies the special case of the clause beginning "unless" in the statement of the linguistic relativity principle at the end of the preceding paragraph. From this condition follows the unanimity of description of the world in the community of modern scientists. But it must be emphasized that "all modern Indo-European-speaking observers" is not the same thing as "all observers." That modern Chinese or Turkish scientists describe the world in the same terms as Western scientists means, of course, only that they have taken over bodily the entire Western system of rationalizations, not that they have corroborated that system from their native posts of observation.

When Semitic, Chinese, Tibetan, or African languages are contrasted with our own, the divergence in analysis of the world becomes more apparent; and, when we bring in the native languages of the Americas, where speech communities for many millenniums have gone their ways independently of each other and of the Old World, the fact that languages dissect nature in many different ways becomes patent. The relativity of all conceptual systems, ours included, and their dependence upon language stand revealed. That American Indians speaking only their native tongues are never called upon to act as scientific observers is in no wise to the point. To exclude the evidence which their languages offer as to what the human mind can do is like expecting botanists to study nothing but food plants and hothouse roses and then tell us what the plant world is like!

Let us consider a few examples. In English we divide most of our words into two classes, which have different grammatical and logical properties. Class 1 we call nouns, e.g., 'house, man'; class 2, verbs, e.g., 'hit, run.' Many words of one class can act secondarily as of the other class, e.g., 'a hit, a run,' or 'to man (the boat),' but, on the primary level, the division between the classes is absolute. Our language thus gives us a bipolar division of nature. But nature herself is not thus polarized. If it be said that 'strike, turn, run,' are verbs because they denote temporary or short-lasting events, i.e., actions, why then is 'fist' a noun? It also is a temporary event. Why are 'lightning, spark, wave, eddy, pulsation, flame, storm, phase, cycle, spasm, noise, emotion' nouns? They are temporary events. If 'man' and 'house' are nouns because they are long-lasting and stable events, i.e., things, what then are 'keep, adhere, extend, project, continue, persist, grow, dwell,' and so on doing among the verbs? If it be objected that 'possess, adhere' are verbs because they are stable relationships rather than stable percepts, why then should 'equilibrium, pressure, current, peace, group, nation, society, tribe, sister,' or any kinship term be among the nouns? It will be found that an "event" to us means "what our language classes as a verb" or something analogized therefrom. And it will be found that it is not possible to define 'event, thing, object, relationship,' and so on, from nature, but that to define them always involves a circuitous return to the grammatical categories of the definer's language.

In the Hopi language, 'lightning, wave, flame, meteor, puff of smoke, pulsation' are verbs—events of necessarily brief duration cannot be anything but verbs. 'Cloud' and 'storm' are at about the lower limit of duration for nouns. Hopi, you see, actually has a classification of events (or linguistic isolates) by duration type, something strange to our modes of thought. On the other hand, in Nootka, a language of Vancouver Island, all words seem to us to be verbs, but really there are no classes 1 and 2; we have, as it were, a monistic view of nature that gives us only one class of word for all kinds of events. 'A house occurs' or 'it houses' is the way of saying 'house,' exactly like 'a flame occurs' or 'it burns.' These terms seem to us like verbs because they are inflected for durational and temporal nuances, so that the suffixes of the word for house event make it mean long-lasting house, temporary house, future house, house that used to be, what started out to be a house, and so on.

Hopi has one noun that covers every thing or being that flies, with the exception of birds, which class is denoted by another noun. The former noun may be said to denote the class (FC–B)—flying class minus bird. The Hopi actually call insect, airplane, and aviator all by the same word, and feel no difficulty about it. The situation, of course, decides any possible confusion among very disparate members of a broad linguistic class, such as this class (FC–B). This class seems to us too large and inclusive, but so would our class 'snow' to an Eskimo. We have the same word for falling snow, snow on the ground, snow packed hard like ice, slushy snow, wind-driven flying snow—whatever the situation may be. To an Eskimo, this all-inclusive word would be almost unthinkable; he would say that falling snow, slushy snow, and so on, are sensuously and operationally different, different things to contend with; he uses different words for them and for other kinds of snow. The Aztecs go even farther than we in the opposite direction, with 'cold,' 'ice,' and 'snow' all represented by the same basic word with different terminations; 'ice' is the noun form; 'cold,' the adjectival form; and for 'snow,' "ice mist."

What surprises most is to find that various grand generalizations of the Western world, such as time, velocity, and matter, are not essential to the construction of a consistent picture of the universe. The psychic experiences that we class under these headings are, of course, not destroyed; rather, categories derived from other kinds of experiences take over the rulership of the cosmology and seem to function just as well. Hopi may be called a timeless language. It recognizes psychological time, which is much like Bergson's "duration," but this "time" is quite unlike the mathematical time, T, used by our physicists. Among the peculiar properties of Hopi time are that it varies with each observer, does not permit of simultaneity, and has zero dimensions; i.e., it cannot be given a number greater than one. The Hopi do not say, "I stayed five days," but "I left on the fifth day." A word referring to this kind of time, like the word day, can have no plural. The puzzle picture (Fig.

11, page 213) will give mental exercise to anyone who would like to figure out how the Hopi verb gets along without tenses. Actually, the only practical use of our tenses, in one-verb sentences, is to distinguish among five typical situations, which are symbolized in the picture. The timeless Hopi verb does not distinguish between the present, past, and future of the event itself but must always indicate what type of validity the SPEAKER intends the statement to have: (a) report of an event (situations 1, 2, 3 in the picture); (b) expectation of an event (situation 4); (c) generalization or law about events (situation 5). Situation 1, where the speaker and listener are in contact with the same objective field, is divided by our language into the two conditions, 1a and 1b, which it calls present and past, respectively. This division is unnecessary for a language which assures one that the statement is a report.

Hopi grammar, by means of its forms called aspects and modes, also makes it easy to distinguish among momentary, continued, and repeated occurrences, and to indicate the actual sequence of reported events. Thus the universe can be described without recourse to a concept of dimensional time. How would a physics constructed along these lines work, with no T (time) in its equations? Perfectly, as far as I can see, though of course it would require different ideology and perhaps different mathematics. Of course V (velocity) would have to go too. The Hopi language has no word really equivalent to our 'speed' or 'rapid.' What translates these terms is usually a word meaning intense or very, accompanying any verb of motion. Here is a clue to the nature of our new physics. We may have to introduce a new term I, intensity. Every thing and event will have an I, whether we regard the thing or event as moving or as just enduring or being. Perhaps the I of an electric charge will turn out to be its voltage, or potential. We shall use clocks to measure some intensities, or, rather, some RELATIVE intensities, for the absolute intensity of anything will be meaningless. Our old friend acceleration will still be there but doubtless under a new name. We shall perhaps call it V, meaning not velocity but variation. Perhaps all growths and accumulations will be regarded as V's. We should not have the concept of rate in the temporal sense, since, like velocity, rate introduces a mathematical and linguistic time. Of course we know that all measurements are ratios, but the measurements of intensities made by comparison with the standard intensity of a clock or a planet we do not treat as ratios, any more than we so treat a distance made by comparison with a yardstick.

A scientist from another culture that used time and velocity would have great difficulty in getting us to understand these concepts. We should talk about the intensity of a chemical reaction; he would speak of its velocity or its rate, which words we should at first think were simply words for intensity in his language. Likewise, he at first would think that intensity was simply our own word for velocity. At first we should agree, later we should begin to disagree, and it might dawn upon both sides that different systems of rationalization were being used. He would find it very hard to make us understand what he really meant by velocity of a chemical reaction. We should have no words that would fit. He would try to explain it by likening it to a running horse, to the difference between a good horse and a lazy horse. We should try to show him, with a superior laugh, that his analogy also was a matter of different intensities, aside from which there was little similarity between a horse and a chemical reaction in a beaker. We should point out that a running horse is moving relative to the ground, whereas the material in the beaker is at rest.

One significant contribution to science from the linguistic point of view may be the greater development of our sense of perspective. We shall no longer be able to see a few recent dialects of the Indo-European family, and the rationalizing techniques elaborated from their patterns, as the apex of the evolution of the human mind, nor their present wide spread as due to any survival from fitness or to anything but a few events of history—events that could be called fortunate only from the parochial point of view of the favored parties. They, and our own thought processes with them, can no longer be envisioned as spanning the gamut of reason and knowledge but only as one constellation in a galactic expanse. A fair realization of the incredible degree of diversity of linguistic system that ranges over the globe leaves one with an inescapable feeling that the human spirit is inconceivably old; that the few thousand years of history covered by our written records are no more than the thickness of a pencil mark on the scale that measures our past experience on this planet; that the events of these recent millenniums spell nothing in any evolutionary wise, that the race has taken no sudden spurt, achieved no commanding synthesis during recent millenniums, but has only played a little with a few of the linguistic formulations and views of nature bequeathed from an inexpressibly longer past. Yet neither this feeling nor the sense of precarious dependence of all we know upon linguistic tools which themselves are largely unknown need be discouraging to science but should, rather, foster that humility which accompanies the true scientific spirit, and thus forbid that arrogance of the mind which hinders real scientific curiosity and detachment.

"The Level of Water." Reprinted from The Child's Conception of Physical Causality by Jean Piaget, pages 164-174, by permission of the Humanities Press, Inc., Totowa, New Jersey: Littlefield, Adams, and Co., 1966.

CHAPTER VII

THE LEVEL OF WATER

By JEAN PIAGET

NOTHING is better suited to throw light on the dynamic significance attributed to weight by very young children, and on the difficulty which they have in taking volume into account, than the immersion of a pebble in a glass of water. ·

The problem we are going to set the children is extremely simple, and that is its great advantage over the problem of the floating boats. The child is shown a glass three-quarters full of water and a pebble. We say: "I am going to put this pebble into the water, right in. What will happen? What will the water do?" If the child does not immediately say: "The water will go up," we add: "Will the water stay at the same place or not?" Once the child has given his answer, the experiment is done, the child is asked to note that the level of the water has risen, and is asked to explain this phenomenon. The youngest children always answer that the water has risen because the pebble is heavy and weighs on it. The child is then given a much bulkier, but lighter object, and is asked whether it, too, will make the water rise, and why. The experiment can be varied with nails, with shot, with wood, etc.

It goes without saying that this interrogatory must be made before the children have been questioned on the subject of the boats, so as to avoid perseveration. The children whose answers we are going to quote were questioned first about the present subject and only afterwards about the boats.

Three stages may be distinguished in the explanations

144

of the displacement of the water-level. During a first stage (under 7-8), the water is supposed to rise because the pebble is heavy. From the point of view of prediction, the child is consistent : a large pebble will make the water rise less high than a collection of very heavy grains of lead. During a second stage (7 to 9 years), the prediction is correct : the child knows that the submerged bodies will make the water rise in proportion to their bulk. But in spite of this correct prediction the child continues, oblivious to the constant self-contradictions which he becomes involved in, to explain the phenomenon by weight and not by volume : the submerged body, he says, makes the water rise in proportion to its weight. Finally, during a third stage (from 10-11 onwards), the correct explanation is found.

§ I. FIRST STAGE : THE WATER RISES BECAUSE OF THE WEIGHT OF THE SUBMERGED BODY.—Let us start with an observation taken from ordinary life. A little girl of 9, in her own home, who has never been questioned on the present subject, is on the point of putting a large bunch of flowers into a vase full of water. She is stopped : "Take care ! It will run over ! " The child answers : " *No, because it isn't heavy.*" Thus to her mind it is not the volume of the body that matters but simply its weight : the bunch of flowers, not being heavy, can enter the water without exercising pressure and consequently without raising its level.

This little piece of everyday observation will be found to tally with the more general results which characterise our first stage. Children of this stage think that submerged bodies make the water rise in virtue of their weight : a small, heavy body will bring about a greater rise in the level than a large body of lesser weight.

What is the meaning of this statement ? At first sight, it looks as though the child simply confused volume with weight, and designated volume by the word " heavy ". It might also appear as though, in the child's eyes, heavy bodies made the level rise owing to their bulk. Above all, it looks as though, when the child speaks of weight, all he wanted to say was that heavy objects went completely inside the water and in that way raised its level. In point of fact, children of this stage do confuse weight and volume. But these children are not thinking of the displaced volume when they say that heavy objects make the water rise because they are heavy. They think that the submerged body exercises a continuous pressure in the water and thus raises its level, not because it occupies space, but because it sets up a current which runs from the bottom upwards like a wave. Here are some examples :

KEN (7) predicts that the water will not rise. We slip in the pebble, but without making a ripple on the surface (which is important) : " Will that make the water rise ? —*Yes.*—Why ?—*Because it hit.*—Where did it hit ?— *At the bottom.*—Why does the water rise ?—*Because it hit at the bottom.*—But why does that make it rise ?—*Because there is a wind* ! [It will be remembered that for many children a " current " of water is identified with a " current " of air, in other words the wind.] (See Chap. IV, § 2).—If I put this pebble [a larger pebble] very gently ?—*It will rise.*—Why ?—*Because it hits.* [We make the experiment.]—Why does it rise ?—*It makes a wind.*—Where ?—*In the water.*" " Which will make the water rise higher, this pebble or this one ?—*This one* [the larger].—Why ?—*It's bigger.*—Then why will it make the water rise higher ?—*Because it is stronger.*" It will be seen that the thought of volume does not enter in, and that the child is concerned solely with factors of a dynamic order.

ZWA (8 ; 3) immediately foresees the phenomenon. The water " *will go up because the pebble made it go up by falling.*" It seems, then, that Zwa perceives clearly the rôle of volume. Nothing of the kind : " Then why does the pebble make it rise ?—*It rises because the pebble is heavy.* [We make the experiment with a small pebble.]— *It didn't go up much because the pebble went gently.*—What must you do to make it go up ?—*You must put the pebble in hard.* [We make the experiment, but without throwing the pebble or producing ripples.]—*The water has risen a lot because the pebble went in a little harder than before.*— Why has the water risen ?—*Because it went down ; it went rather hard and the water has risen.*—But why does the pebble make the water rise ?—*Because the pebble is a bit stronger than the water. So that makes the water lift up.*— Why is the pebble strong ?—*Because the pebble is big and the water is light.*" [" Big " here evidently means " heavy " or " condensed."] In front of Zwa we place a piece of wood on the water and the water rises moderately. We then show to Zwa a small pebble and a piece of wood, the pebble is heavier, although smaller, than the wood : " Which will make the water rise most ?—*The pebble because it is stronger.*—Why will it make it rise more ?— *Because the pebble is smaller, but bigger* [=condensed], *heavier.*" Zwa then discovers, to his astonishment, that it is the wood which produces the largest rise in the level of the water. " *The wood made the most because it stayed on the top, and then it's rather heavy, and then as soon as it touches the water it is strong* [!] *and that makes it rise.*" We then show to Zwa some nails and a pebble of the same weight ; the nails have a much smaller volume. Zwa says : " *They are both the same heaviness, they will both make it rise the same.*" It will be seen that Zwa never thinks of volume. He only speaks of weight, strength, and the current.

MAI (8½) not only does not foresee that the water will rise, but also predicts that the level will sink : " *It will go down.—Why ?—Because the pebble is heavy.—And if I put two pebbles ?—It will go down.—More* or less than with one pebble.*?—More.*" Thus Mai is of opinion that the pebble compresses the water. We make the experiment. Mai cries : " *Ooh ! It's gone up !—Why ?—Because the pebbles are heavy.*" We insist : " Why does the pebble make the water rise ?—*Because the pebble is heavy, and then it is hard, and then . . . when it's at the bottom it makes little balls* [air bubbles—but Mai takes them to be bubbles of water], *and then they go up and come to bits.*" In other words, the water comes out of the bubbles and raises the level. " Why are there these balls ?—*Because the pebble is at the bottom. . . .*—Why does that make balls ?—*Because it is heavy.*"

PERE (11) : " Will the water rise or not ?—*It will rise just a little because the pebble is heavy and then that makes the water rise.—Why ?—Because it has a lot of weight.*" Peré predicts that a piece of wood will not make the water rise. He is afterwards very astonished to observe the contrary. " Why has it risen ?—*Because the water is not stronger than the wood.*—Why does it rise then ?—*Because there is no air inside* [Peré means to say that the water, having no current, is unable to resist the pressure of the wood], *because that always makes it rise when you put things inside. That makes it rise.*—Why ?—*Because it makes a weight.*—Is this piece of wood heavy ?—*No, but there's already a stone inside. That makes it rise.*" In other words, the stone continues to make the water rise, when in addition a piece of wood is placed on the top. We show to Peré a little bag containing small shot, and a large pebble, the shot being the heavier : Peré predicts that the bag will raise the level much higher than the pebble " *because the bag is heavier.*"

MIE (10) : " Will the water rise if I put this pebble in ? —*Yes, a little, because it is heavy at the bottom.*"

WENG (8½) : " *It weighs on the water, that makes it go up.*"

MOUL (8½) : " *It will overflow* [although the tumbler is only three-quarters full and the pebble is not large].— —Why ?—*Because it is heavy.* [We make the experiment.]—*The pebble has fallen and the water stays inside !—* Why did it not overflow ?—*Because the pebble is not heavy enough.*" So Moul takes no account of volume.

GESS (9½) predicts that the water will rise. " Why ?— *Because the pebble is heavy.*—Why does that make the water rise ?—*Because the water is light and when you put a pebble in it is a bit heavier, so it rises.*" Thus the water is supposed to expand in becoming heavier, or under pressure.

CESS (12) says that a small pebble will make the water rise higher than a large piece of wood " *because the pebble is bigger, not bigger, but stronger, and that will make the water rise higher.*"

These cases are all clear and definite. But there is an experiment which shows conclusively that when the child explains the rise in the water-level by the weight of the submerged body, he is really thinking of an upward thrust or current and not in any way of volume. When the child says that the water rises because the pebble " knocks the bottom," we simply ask him whether a pebble held by a thread halfway down the column of water would also raise the level. Children of this stage generally answer that it would not, for a pebble that is held by something no longer weighs on the water. Here is an example :

GEN (6 ; 8) predicts that the pebble will make the water rise " *because it is heavy in the water. So it rises.*" " You see the pebble hanging on this thread : if I put it in the water as far as this [half-way], will it make the water rise ? —*No, because it is not heavy enough.*" It is, however, the same pebble.

Weight, according to Gen, is the capacity for exercising a real activity : it is the action of expanding the water, of " making a current." This, it would seem, is the belief universally held at this stage.

In a word, although nearly all the children were able to foretell that the water-level would rise with the immersion of the pebble, not one of them brought the fact of volume into his explanation. They all appealed to weight, with the idea that the pressure of the pebble produces a current, bubbles, an expansion of the water, and so on.

Tendencies of two sorts existing in the child's mind serve to explain the phenomenon. In our analysis of the suspension of clouds and of boats we saw how much more dynamic than mechanistic are the schemas existing in the child's mind. The same tendency is at work here. The explanations we have just quoted even furnish an additional clue to the understanding of those curious statements made by the children about clouds staying in the air or boats floating in the water because they were heavy and big. This means that clouds and boats, by weighing on the air or the water, liberate a current sufficiently strong to keep them in place. This is probably a fresh version of the schema of " reaction of the surrounding medium " which was studied in an earlier chapter.

The very constancy of the phenomena of this stage is due also to a tendency which makes all very young children identify weight and volume. To children who regard weight as always, or nearly. always, proportional to volume, heavy objects are sure to be those that raise the level of the water. Their observation of fact is roughly correct. But their interpretation is defective, owing to the dynamic turn of mind of which we have been reminding the reader, and which entirely neglects the factor of weight in favour of the factor of volume.

§ 2. SECOND AND THIRD STAGES : THE RÔLE OF VOLUME IS FELT AND MADE EXPLICIT.—Nothing is so well designed to show the child's difficulty in overcoming his spontaneous dynamism than the existence of this second stage. For during this period the child bases all his predictions on volume, and he will roundly declare that a bulky but light piece of wood will make the water rise higher than a heavy but small pebble. The interesting thing, however, is that the child is not conscious of this choice : he persists in explaining the rise by appealing only to the weight of the submerged body, ignoring the while the incessant contradictions in which he is involved by systematically taking up this attitude. A circumstance of this kind is of the utmost interest in the psychology of childish reasoning. The difficulty of bringing all the relevant factors into consciousness, the difficulty of performing logical demonstration, the part played by motor intelligence as opposed to that of conceptual or verbal intelligence — all these important questions converge on this one point of physics.

Here are some examples :

MEY (10 ; 8) : The water rises when the pebble is put in " *because it weighs on it.*—Why has the pebble made the water rise ?—*Because it is heavy.*—Which is the heavier, this wood [bulky] or this pebble [small] ?—*The pebble is heavier.*—Which will make the water rise higher ? —*The wood.*—Why ?—*Because it is lighter.*" We again show a small pebble and a large piece of wood : " Which will make the water rise higher ?—*The wood.*—Why ?— *Because it is lighter.*—Is it because it is light that it makes the water rise ?—*Yes.*—Why ?—*Because it* [the wood] *weighs a very little. It weighs and so that makes the water go up.*—But why does the wood make the water rise more than the pebble ?—*Because the wood is lighter than the Because they are heavy.*" A moment later, we place a certain quantity of aluminium in the water and the water overflows, conformably to Müll's prediction. " Why did that make the water rise ?—*Because it's light.*"

Ro (6½, very advanced in every way) also predicts that the pieces of wood will make the water rise better than the pebble, but he adds that it is because they are " heavy." However, he has weighed them and observed that they were lighter than the pebbles. At a given moment Ro speaks of volume, and when asked what will happen says, " *You must see if they are the same size,*" but at the same time he weighs the objects to be compared as though the size were estimated by the weight.

Thus even in children who are on the threshold of the third stage, like Müll and Ro (though incidentally Müll and Ro talk about volume), explanation by weight is still extraordinarily persistent.

Let us turn to the analysis of the third stage which begins about the age of 9-10. This stage is marked by the appearance of explanation by volume. Thus explanation links up with prediction.

Here are some examples :

BIZ (10 ; 3) : " What will the water do when I put this pebble in it ?—*It will rise.*—Why ?—*Because the pebble takes up space.*—If I put this wood, what will happen ?—*It will lie on the water.*—And what will the water do ?—*It will rise because the wood also takes up space.*—Which is the heavier, this pebble [small] or this wood [large] ?—*The pebble.*—Which will make the water rise the most ?—*The pebble takes up less space, it will make the water rise less.*"

PERN (10 ; 11) : " *The water will rise because the pebble will take up room.*' The wood will make the water rise more than the pebble does " *because it takes up more space.*"

KIM (11) : " *The water will rise because the pebble takes up space at the bottom.*"

MEN (12) : The water rises " *because the pebble is rather heavy, so it goes a little to the bottom, and that takes up space.*—And the wood ?—*The water rises a little because you can see a little wood go down, and that takes up space.*—Which will make the water rise most, the wood, which

Because they are heavy." A moment later, we place a certain quantity of aluminium in the water and the water overflows, conformably to Müll's prediction. " Why did that make the water rise ?—*Because it's light.*"

Ro (6½, very advanced in every way) also predicts that the pieces of wood will make the water rise better than the pebble, but he adds that it is because they are " heavy." However, he has weighed them and observed that they were lighter than the pebbles. At a given moment Ro speaks of volume, and when asked what will

happen says, " *You must see if they are the same size,*"
but at the same time he weighs the objects to be compared as though the size were estimated by the weight.

Thus even in children who are on the threshold of the
third stage, like Müll and Ro (though incidentally Müll
and Ro talk about volume), explanation by weight is still
extraordinarily persistent.

Let us turn to the analysis of the third stage which
begins about the age of 9-10. This stage is marked by
the appearance of explanation by volume. Thus explanation links up with prediction.

Here are some examples :

BIZ (10 ; 3) : " What will the water do when I put this
pebble in it ?—*It will rise.*—Why ?—*Because the pebble
takes up space.*—If I put this wood, what will happen ?—
It will lie on the water.—And what will the water do ?—*It
will rise because the wood also takes up space.*—Which is the
heavier, this pebble [small] or this wood [large] ?—*The
pebble.*—Which will make the water rise the most ?—*The
pebble takes up less space, it will make the water rise less.*"

PERN (10 ; 11) : " *The water will rise because the pebble
will take up room.*' The wood will make the water rise
more than the pebble does " *because it takes up more
space.*"

KIM (11) : " *The water will rise because the pebble takes
up space at the bottom.*"

MEN (12) : The water rises " *because the pebble is rather
heavy, so it goes a little to the bottom, and that takes up
space.*—And the wood ?—*The water rises a little because
you can see a little wood go down, and that takes up space.*—
Which will make the water rise most, the wood, which
is light, or this pebble, which is heavy ?—*The wood, because
it is bigger.*"

One is almost surprised at the simplicity of these
answers after the complexity of those of the earlier stages.
Weight no longer comes in at all, except to explain how
the object becomes submerged. And at no point is the
submerged body supposed to produce a current from below
upwards. The displacement of volume alone explains
why the level of the water has risen.

§ 3. CONCLUSIONS.—From the point of view of causality,
the foregoing explanations evolve according to a very
definite law which is the same as in the case of the floating
boats : the child proceeds from dynamic to mechanical
causality. According to the youngest children, the pebble
is active : it makes " wind ", " bubbles ", " a current ",

etc. According to the older ones, weight accounts only for the immersion of the body, and the rise of the water-level is due to the displacement of volume.

Such a principle of evolution as this accounts not only for the curious explanations, according to which clouds and boats keep themselves up by means of their weight, but also for the difficulty which children find in taking account of the volume of a floating body in their explanations.

Cross-Cultural Teaching of Science

Study of the intellectual environment in which children live may lead to better science teaching.

Francis E. Dart and Panna Lal Pradhan

A major theme of our age is the development of science and technology in societies around the world. Leaders in most of the developing countries of Asia, Africa, and South America, where knowledge and utilization of natural resources have remained nearly static over the past several centuries, now recognize that they must move into the era of applied science and technology if provision is to be made for the needs of increasing populations and for improved standards of living (1). Thus, technology and science are emphasized in their development plans and in the assistance they seek and receive from nations such as the United States (2). It is generally assumed that this process of scientific and technological development will require very much less time in Asia, Africa, and South America than it did in Europe and North America, and in fact many countries hope to achieve in one or two generations changes comparable to those that occurred in the West over 2 or 3 centuries. Their hope is based partly upon the availability of capital assistance from the industrial nations and partly upon the ease and rapidity with which knowledge now at hand can be communicated—knowledge which was originally obtained over a long period in a process involving many errors and confusions that will not have to be repeated. In their optimism they largely ignore the profound social and cultural changes that accompanied the Western development and the social and cultural changes that must accompany this new scientific revolution. Frequently it is found that a country whose leaders are determined to introduce rapid change is not ready to adopt modes of thought and organization that are fundamental to an advancing science and technology, and hoped-for results have been slow to materialize. It should be added that unnecessary ambiguity has sometimes resulted from a failure, in discussions such as this one, to distinguish between technology and science (3). In what follows we are concerned specifically with science and with problems associated with its introduction.

The difficulties encountered frequently relate to the very nature of the interaction between Western science and non-Western cultures, an interaction that has received little study in spite of the fact that it lies at the very heart of the development process. Western

153

technology developed out of the Western scientific revolution, which, over the last 3 centuries, has profoundly altered Western man's understanding of, and relation to, nature. The resulting "scientific viewpoint" has become our way of considering reality, and it is so much a part of us that it is taken for granted. The traditional cultures of Asia or Africa, however, are frequently nonscientific—nonrational in their approach to nature—and they do not always provide a ready foundation upon which to build a more scientific view. Of course people of all cultures experience many of the same familiar phenomena of nature and feel that they understand what is real and how knowledge about the real world is to be organized. Interpretations of what is meant by the "real" world, however, vary widely. Major tasks, then, are to determine what constitutes reality for persons of different cultures and to learn how the most meaningful communication about nature can be established among people holding different views of reality. The first of these tasks has been undertaken, with particular attention to science, by Malinowski (4), Hsu (5), and others (6), but our knowledge is far from complete; the second has hardly been touched on, except as it relates to science education within the Western countries.

Science education, in any country, is certainly a systematic and sustained attempt at communication about nature between a scientific and a nonscientific, or a partially scientific, community, and as such it should be particularly sensitive to the attitudes and presuppositions of both the scientist and the student. In fact, however, the teaching of science is often singularly insensitive to the intellectual environment of the students, particularly so in the developing countries, where the science courses usually offered were developed in a foreign country and have undergone little if any modification in the process of export. Why should we suppose that

a program of instruction in botany, say, which is well designed for British children, familiar with an English countryside and English ways of thinking and writing, will prove equally effective for boys and girls in a Malayan village? It is not merely that the plants and their ecology are different in Malaya; more important is the fact that the *children* and *their* ecology are also different.

We are convinced that a study of the intellectual environment in which children live can lead to significant improvements in science teaching and science learning. This is of particular importance, moreover, to the developing countries whose environments are very different from those of the West and whose educational resources are so limited as to make any increase in efficiency very desirable. We discuss in this article an initial effort in that direction, some pilot experiments conducted in Nepal during October and November of 1965.

Experimental Setting and Procedure

Nepal and its people remained effectively isolated from Western intervention or education throughout the entire colonial period of European influence in Asia, the only important exception being the Gurkha mercenary soldiers who returned to their villages after a period of service in the British army, bringing with them the accumulated experiences of several years of travel and contact with foreign places and ideas. Only within the last dozen years has there been any opportunity for Western education or science to reach appreciable numbers of Nepalese communities, and there is now a considerable range in the degree to which they have penetrated into village life and thought. The government is actively supporting the development of education, as to both quantity and quality, through increase in the number of schools, establishment of a modern teachers college, development of a national university, and other measures. These circumstances, together with the practical consideration that one of us (P.L.P.) is a native of the country, with knowledge of the language and customs, dictated our choice of Nepal for our investigations. (It should be added that each of us has spent several years living and working in the other's native country.)

We decided to investigate three widely separated ethnic communities having quite different histories of outside contact: the Newars of the Kathmandu Valley, the Limbus of eastern Nepal, and the Gurungs and Chhetris in the west of Nepal. The sacrifice in depth which this decision entails is justified by a need, at least at this early stage, to determine how widely applicable our conclusions might be. The Newars have, for many centuries, been the principal inhabitants of the Kathmandu Valley, where they have developed a rich artistic and literary heritage. They include many skillful artisans and enterprising merchants within a predominantly agricultural economy. They are conservative, adhering rigidly to an inclusive, self-consistent social and philosophical orthodoxy. The Newar town of Panga, where we worked, is a closely packed unit of narrow paved streets and three-story brick houses surrounded by fertile rice fields. Many of its citizens make frequent trips to Kathmandu, the cosmopolitan capital no more than 8 kilometers away, and some of them work there regularly, yet Panga is in most respects an island which might just as well be 500 kilometers or 5 centuries away. Its people are shy, friendly, and hospitable to strangers, but they express little curiosity about a visitor's thoughts. The town has both primary and secondary schools with trained teachers, but fewer than one-fourth of its school-age children attend school.

In contrast, the Limbu village of Tokma is several days' walk from any motor road, airfield, or city. Its dwellings are scattered widely over a large and fertile hillside, which provides ample income to the inhabitants, all of whom live by farming. There is no school or store or other business establishment in the village itself. The villagers have the reputation, shared by all Limbus, of being proud, quick to anger, and fiercely independent. They have resisted political domination, and they take pride in an independence of mind which resists the importation of orthodoxies from outside. Their religious life combines Hinduism with shamanism and witchcraft, and they maintain unique customs not shared by other Hindus (7). They are reserved and suspicious of visitors but not hostile, adhering carefully to established rules of courtesy.

Finally the Chhetri and Gurung people of Armala Dihi are open, friendly, and relatively poor farmers who will sit for hours asking questions about a stranger's experiences and opinions and about the world he comes from. Many Chhetri men serve as Gurkha soldiers, and nearly every village has one or more members who have returned from such service to live in retirement as respected and influential citizens whose pension payments add significantly to the economic well-being of the village. In Armala Dihi there is a small primary school but no store. There is a good middle school about an hour's walk away. Whereas in Tokma we were the guests of a wealthy Limbu landowner for several days without once being invited to enter his house, our host in Armala Dihi insisted that we move into his own room, sleep in his own bed, share his meals, and use the best of everything he could provide.

We sought information about attitudes toward familiar phenomena of nature, and about the sources of knowledge about nature, through interviews with school-age children, typically 9 to 14 years old, and with adults of an age to be these children's parents.

The interviewing was kept informal and usually involved small groups of three or four individuals at a time. The main content and order of the interviews was held constant for all groups.

The questions were of three types, designed to reveal (i) how the respondents accounted for various commonly experienced phenomena, such as rain, lightning, thunder, fire, and earthquake; (ii) what attitudes the respondents held about the control or manipulation of such phenomena; (iii) what were considered to be the origins of knowledge about nature, and what the accepted criteria of validity of such knowledge were. Typical questions are as follows.

Category i. How do you account for rain? Where does the rainwater originate? What do most people in the village think about rain? What makes an earthquake?

Category ii. How can rainfall be brought about or prevented? Is it appropriate for men to influence the rain? Is there any protection against lightning or thunder?

Category iii. How were these things (about rain, and so on) learned? How does one know if they are true? How might new knowledge about such things be obtained?

In addition, observations were made of the kinds of opportunities available to children for learning and practicing skills of abstraction and manipulation that could later be a help in learning and using science. As a test of ability to represent real situations by means of an abstract model, each respondent was asked to sketch a rough map showing how to get from his house to the school.

For comparison, similar groups of American primary school children, aged 9 to 12, attending the University of Hawaii Elementary School in Honolulu, were interviewed and asked to sketch maps. About half of these children were Caucasian, the others being of Asian and Polynesian origin. All had

been brought up in Hawaii among typically Western surroundings of American games and toys, American magazines and television programs, and all the great diversity of intellectual and physical stimuli to be found in a city such as Honolulu.

The Nature of Phenomena

Throughout the interviews, in both Nepal and Hawaii, our interest was directed not toward the "correctness" of a response, as judged by accepted scientific or other standards, but rather toward the type of the response itself and the relation to nature that it suggests—whether, that is, it suggests an explanation of phenomena that is mechanistic, supernatural, teleological, and so on. If a given statement can be recognized as referring to a certain religious belief, for example, that recognition serves our purpose, and we are not concerned with whether or not the pertinent religious scripture is accurately quoted or even explicitly referred to.

With very few exceptions we were given both a "folk-oriented" or "myth-oriented" and a "school-oriented" explanation of a given phenomenon within a single interview, sometimes by a single individual. Thus, to account for earthquakes, one of a group of four Chhetri boys said, "The earth is supported on the back of a fish. When the fish grows tired it shifts the weight, and this shakes the earth."

All agreed, but another added, "There is fire at the center of the earth. It seeks to escape and sometimes cracks the earth, causing an earthquake." All agreed to this as well.

In a group of Newar school children (four boys and a girl) these statements were given in answer to the same question:

"The earth is supported by four elephants. When one of them shifts the weight to another shoulder an earthquake results."

"There are fire and molten metal inside the earth which try to escape. They may crack or move the rock of the earth, causing an earthquake."

Again all agreed to both statements.

This pattern is repeated again and again:

"The deities break vessels of water in the sky, causing rain."

"The sun evaporates water from the sea, producing vapor which is cooled by the mountains to make clouds and rain."

* * *

"Lightning comes from the bangles of Indra's dancers."

"Lightning comes from the collision of clouds."

* * *

"It rains only in the summer (monsoon) season because we need the rain then. In winter we do not need rain."

"It rains in the summer because the sun is hotter then and causes more evaporation."

* * *

The replies given by Newars, Limbus, and Chhetris are very similar in content, evidently reflecting a common background of mythology and of school curricula, a similarity which is not very surprising, for the three groups, with all their differences, do in fact have a common school system and, in the main, a common religion. More surprising is the fact that each group nearly always gave answers of both the types illustrated above, and that all the members generally accepted both. Of course there is nothing unusual in the thought that a given phenomenon may result from either of two different causes and hence that, in general, each of these causes may be accepted as potentially valid. However, here the two types of "causes" offered appear to be qualitatively so different as to be mutually incompatible, for they suggest conceptually very different ideas of nature. Examination of the replies quoted shows that they do not admit of the type of synthesis which states, "God is the source of rain. He produces rain by causing the heat of the sun to evaporate water from the sea. . . ." It is as difficult for us to accept both as real alternatives as it is to accept them as simultaneously true.

The contradiction is far more apparent to us, however, than to our respondents, who showed no discomfort over it, a fact which should serve to warn the science educator that all is not as it appears on the surface. The philosophies and literature of Asia make great use of paradox, and, to Asians, contradiction may be more intriguing than disturbing. We should not, therefore, discount the possible existence of very deep-rooted patterns of thought not consonant with the "either-or" logic underlying Western science, the logic which makes it so difficult, for instance, for American students to accept the concept of complementarity in modern physics. However, a simpler explanation should also be considered. Much of the teaching and learning in Nepalese schools involves rote memory only and demands very little understanding or conceptualization. Furthermore, many of the teachers and textbook writers belong to the Brahman caste, the priestly class traditionally responsible for the teaching and preservation of orthodox religious beliefs and practices. It is quite possible that, even without any conscious intent on their part, these teachers and textbook writers have taught early "scientific" concepts in such a way as to produce, in combination with a tradition-oriented home environment, a dual view, according to which distinction between myth and science is unnecessary. Even in science teaching in an American elementary school the amount of teleology used is not inconsiderable. In any case, this dual view of nature is a matter that needs to be considered in the planning of revised science teaching methods.

No such duality was evident concerning the control or manipulation of nature (which was always considered

157

appropriate although not always possible). To questions such as, How can rainfall be brought about or prevented? or, Is there any protection against thunder (lightning)?, a single type of reply was always given. Usually control of such natural phenomena is expected to follow from a religious ritual in which it is made explicit that actual control is at the will of a deity who may not always respond. Thus control is uncertain. In some instances, as when the farmers want hail deflected away from crops, the resort is to magic or charms performed by special persons and not associated with a religious ceremony. Charms too may fail, and all such procedures remain ambiguous enough in principle to make convincing empirical tests of their validity hard to manage. Of course there are many common and well-understood technological manipulations of nature which are taken for granted and explained in operational terms. Such is the case, for example, with irrigation, the cooking of food, or the firing of clay vessels.

In no single instance did a member of any one of the groups of Honolulu school children manifest a comparable duality of viewpoint. The explanations offered in answer to the same questions about rain, lightning, and so on were not always factually correct, but they were always "scientific" in concept and usually mechanical. Lightning is produced "when two clouds collide"; the heat of the sun evaporates and "lifts" water up to make rain; parts of the earth "shift," causing earthquakes, and so on. On two occasions the Biblical story of God erecting the rainbow as a promise to the children of Israel was mentioned, and each time the respondent spontaneously pointed out that this is a different kind of statement and not an explanation of rainbows.

None of the members of the comparison groups in Honolulu believed that control or manipulation of natural phenomena of the kinds under consideration was achievable through magical or religious practices; they considered such control either achievable through technological procedures or impossible. Many, but not all, of the individuals who said control was impossible suggested that it might in time become possible. Sometimes procedures were described which are not used and would not work, such as the use of lightning rods to convert lightning into ordinary house current, but even these procedures were always presented as scientific, technological processes without any occult or supernatural element.

The Nature of Knowledge

When our Nepalese respondents were asked to give the source of their knowledge about nature they invariably said that it came "from books" and "from old people." When we asked how the old people found out or how knowledge got into the books they told us it came from earlier generations of "old people" or from other books. When we pressed for some ultimate source, most of our respondents said that these things had always been known, although a few of them referred to legends telling how some particular skill, such as fire building, was given to men by the deities. One Chhetri student suggested that some knowledge might have been obtained by "accidental observations."

We went on to inquire how knowledge hitherto unknown to anyone might be acquired or how it might be sought. We were always told that such new knowledge is not to be expected. Even when we pushed this question so far as to call attention to such "new" discoveries as space travel or transistor radios, which all Nepalese know about, it was held that such things were always known by someone, or else that these are merely new applications of old knowledge. One very tentative exception was offered by a Limbu boy, who suggested that really new knowledge might sometimes come through dreams. We find it hard to believe that more probing would not reveal other exceptions, yet the predominant view is one that pictures human knowledge about nature as a closed body, rarely if ever capable of extension, which is passed down from teacher to student and from generation to generation. Its source is authority, not observation. In fact, experiment or observation was never directly suggested to us as an appropriate or trustworthy criterion of the validity of a statement, or as its source. When one of us stated that a book, after all, is only a more permanent record of someone's observations, the idea was treated as novel and faintly suspect. Given this concept of knowledge, it is no surprise that the schools rely heavily upon rote memory. Memorizing would seem to be the easiest and most efficient way to deal with a closed and limited body of unvarying facts. There are also other well-known and frequently, criticized forces embedded within the formal educational system which strongly reinforce this natural tendency.

It should not be thought, of course, that it is only in Nepal or Asia that students try to learn science by memorizing. Our comparison group in Honolulu showed evidence of considerable, though more limited, reliance on the memorizing of facts given in a book or stated by a teacher. However, members of the Honolulu group all stated that the knowledge originated in observation and experiment, and they believed that new knowledge not only can be obtained but continuously is being obtained.

Use of Abstractions

Science as the scientist thinks of it and would like to see it taught consists, not of a body of more or less isolated facts to be memorized, but of a system of empirically verifiable relationships between more or less abstract concepts. While the concepts are derived from real phenomena, the relationships of science relate concepts, not real objects, and the theories of science are built around "models" which portray in abstract, often mathematical, terms a selectively idealized representation of real phenomena. It is essential for the science student to learn to be at home, at some level of sophistication, with this process, which must surely appear even to the Western layman to be extraordinarily indirect. Much attention is given to this in the recently developed or improved science courses in the United States, which go to great lengths to give students systematic training and practice in skills of abstraction and inference while striving to maintain contact with the real world by subjecting conclusions to observational verification. Of course, informal learning plays a part in this process. The toys children manipulate, the games they play, the activities of the adults they watch and imitate, the conversations they listen to all contribute to the attitudes and skills they develop (8). In everything the child does in school there is an echo of his environment at home.

How much more difficult science must be, then, for a child who lives in a Nepalese village or small town, immersed in a very different environment with its own pervasive non-Western influences. Here he lives close to nature in a direct, particularistic relation of planting and harvesting, with little or no abstraction and little need to generalize. He does not play with mechanical toys or build mechanical models; he plays or watches games of skill or chance but knows little about games of strategy; he rarely sees a book in his home; he rarely has occasion or opportunity to deal with derived or inferred properties or concepts. Certainly his society, or any society, contains a great many abstractions, ranging from spoken or written language all the way to a very complex religious cos-

159

mology, but these are not all particularly useful in preparing the way for science, which wants to hold to a rather special and verifiable relation to nature. Thus, for example, every Nepalese child will be familiar with abstract representations of certain Hindu or Buddhist deities and heroes of religious myths and legends, yet these are not subject to direct or even indirect observational verification, after the manner of science, and they may not be conducive to a scientific approach.

A thorough analysis of the informal intellectual environment, even in one of the groups we visited, would be a major undertaking which we did not have the resources to undertake. Yet we did want to include some tentative assessment of the effects of informal learning as it might bear upon science learning. For this purpose we asked our subjects to sketch rough free-hand maps showing how to get "from your house to the school" (or to some other well-known local landmark). A map is a fairly simple, yet typical, example of a scientific model. It preserves a verifiable 1-to-1 relation to reality and yet it is an abstraction, useful for what is omitted no less than for what is included. Mapping allows for great variety in the way a given reality is represented, and the relationships and inferences derived from a map, while not totally unrelated to reality, nevertheless actually refer to the model and not to the real world. We believe that the maps which children (9) or adults draw to represent a well-known route or neighborhood will reveal with some accuracy their readiness to understand and use other scientific abstractions.

The "maps" we obtained from the Nepalese respondents are all very similar to each other and to the example shown in Fig. 1. Always they include a recognizable *picture* of "my house" and of "the school," the two being connected by a line which seems to denote *the process of going* from one

to the other, not the spatial relationship of one to the other. Thus, the two buildings represented in Fig. 1 are not in fact on the same street or path, being separated by several street intersections and other landmarks, none of which appear on the map. In contrast we show (Fig. 2) a map typical of those drawn by American children in response to the same instructions. Here both house and school are represented by abstract symbols, not pictures, and there is a clear effort to show spatial relationships and to provide needed spatial clues. The propensity of the Nepalese for making maps (whether verbal or graphic) which are *sequential* rather than spatial constructs is not limited to school children. In a land of foot trails, where literacy is too low to justify the use of signs, this propensity has been a source of consternation to more than a few travelers of Western upbringing! We, too, in reply to our inquiries as we traveled, were given instructions or "maps" which, like a string of beads, list in correct sequence the places we should pass through without giving any clue as to distances, trail intersections, changes of directions, and so on. Our interest is not in the accuracy or potential usefulness of this different kind of model but in the light it may shed on a way of thinking which may extend far beyond mere map making. The villagers use no other kind of map; they do not use drawings in constructing a building or a piece of furniture—in fact they hardly use drawings or spatial representation at all (except for records of land ownership, which does not change very frequently), and the lack of spatial models may be very natural. One wonders, however, whether the science teacher will have this in mind when he presents a model of a molecule or the solar system.

Variations between Groups

Our observations were the same for

all three of the Nepalese groups we visited. No doubt this is partly due to the rather gross nature of the study, which may bring out only the most obvious and general conclusions where a finer-grained, more extensive study could be expected also to reveal characteristics that pertain to one group or v. From the standpoint of science education, however, it is useful to start with observations that lead to widely applicable recommendations, and this has been a factor in the design of our study. Certain differences between groups are suggested by the results of our interviews, however, and we mention below two that might well be of further interest.

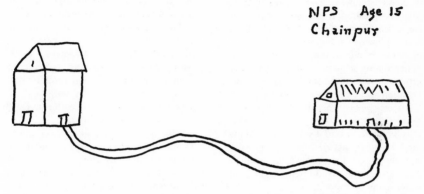

Fig. 1. Map drawn by a 15-year-old Limbu boy to show the way from his house to the school. In fact, the house and school are not on the same street or path.

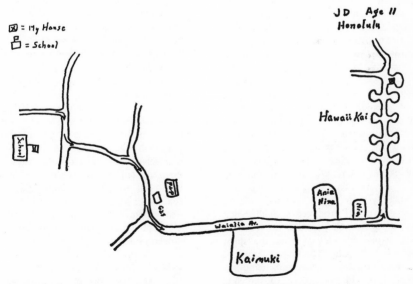

Fig. 2. Map drawn by an 11-year-old American boy to show the way from his house to the school. Note the wealth of spatial and directional clues.

161

Although observation was practically never suggested explicitly by any group as a reliable source of knowledge, we did find many indications among Limbu respondents that observation of nature plays an important role in shaping their attitudes, and also that they feel a need for observational support of theories to an extent not found in either of the other groups. Thus, whereas our Newar respondents always had a firm answer to every question, it was not uncommon for a Limbu to give a reply and then add, "it seems so, but we are not sure," or even to admit that he had no explanation of a given phenomenon. All groups described the rainbow as a manifestation which draws water up from lakes or rivers into the sky. However, more than one of the Limbu respondents pointed out to us that a rainbow may be seen in waterfalls when the sun shines on them, "and so it must have something to do with light and water spray or vapor."

Newar respondents always attributed lightning damage to a particular variety of thunder (*10*), literally "ax-thunder" (we refer to it here as a "thunderbolt"). They describe it as a material object, shaped like an ax, which falls from the sky during a thunderstorm and returns to the sky again, splitting or smashing things on its way. They offer no evidence to support this description, except to refer to the knowledge of old people, if evidence is demanded. When we asked Limbus about thunder and thunderbolts we were told essentially the same thing, with the added information that "thunderbolts" are black in color. We did not question this explanation nor ask for any evidence, yet in a few minutes one of the informants produced a small, black stone artifact shaped like an ax-head. He explained that it was a "spent thunderbolt" which had been damaged when it struck the earth and, being imperfect, could not return to the sky. Further inquiry indicates that such artifacts are occasionally found on the ground or in rice paddies. They are not made from locally available stone and are not believed to be of human origin but are always taken to be "thunderbolts." They fit the mythological concept of "ax-thunder" closely enough to make identification of them in this way very natural, provided a need for observational support of the theoretical concept is felt. Obviously our informant felt that, in our minds too, this evidence would support his explanation.

This experience was duplicated in every essential detail in the community of Armala Dihi 300 kilometers to the west, where we were again presented with a stone-age "thunderbolt"! Other evidence of a need for empirical observation was less strong, however, in Armala Dihi than in Tokma.

Finally, there are differences among these three groups in the degree of stress produced as a result of new ideas from the West. All groups and all individual respondents know that change is coming rapidly to Nepal and that many new ideas are reaching the young people through the rapidly expanding school system. Among the Chhetris of Armala Dihi this is evidently a source of much personal and intergroup tension. The adults welcome schools and urge or require their children to attend them, yet they are engaged with the children in a continuing and sometimes strenuous dialogue in defense of the old ideas and cosmology—a dialogue in which both adults and youths did their best to enlist our participation. Although the adults clearly identify themselves with the old order, they are intensely curious about the new. They seem not to tire of asking questions that range widely over a world of ideas just visible over a still-distant horizon. None of our respondents had served as Gurkha soldiers, but, as we have pointed out, the region includes many individuals who have. In contrast, the Newars of Panga, certainly the most conservative of the three communities, show very little curiosity about the world outside Panga and evidence very little concern about what their children may be learning. The Limbus, too, showed little concern about foreign ideas, which they know about but do not perceive as threatening. They seemed to show a similar lack of concern about a number of introduced Hindu rituals which are practiced and yet disclaimed as foreign and of no value. Such patterns of receptivity or rejection of change and the stresses they lead to in individuals or families are, of course, too complicated to be treated here. Yet they are important to the processes of development and education, and they deserve careful study in considerable depth.

Implications for Science Education

The foregoing observations suggest changes in method or content which might lead to easier and more economical, as well as more effective, teaching of science in Nepal. We believe that some of these changes will be found to apply more or less well in many other developing countries, and we hope that similar, or more refined, studies will be undertaken which will extend and, hopefully, corroborate our conclusions. For the present we shall limit ourselves to a few relatively clearly indicated changes that could be introduced on a pilot scale.

It is clear that the school-age boys and girls among our subjects do not have the attitudes about nature and learning that are most conducive to an understanding of science. Clearly, too, they have not developed much skill in abstract representation, measurement, and so on—skills which contribute not only to scientific experimentation but to conceptualization as well. It is now well accepted that such attitudes and skills can best be developed relatively early in a child's school experience, well before the introduction of formal subject-matter courses like chemistry or physics. We believe that a program of pre-science instruction in the elementary grades, similar to programs now under development in the United States (11), will be both possible and very desirable in Nepal and probably in other developing countries. This instruction may well follow the general guidelines that have been laid down for the American efforts, but it will have to be adapted to conform closely to the particular environment, needs, and available resources of the country and community where it is used, and it should start with the questions children ask there. The project will involve program design and teacher training but no difficult economic problems pertaining to equipment or supplies, for local "phenomena" for observation are best and are abundant; "laboratory material" will consist of leaves and pebbles, sunshine and seeds; and equipment will consist of pieces of bamboo, locally available utensils, and so on. Such a program can certainly present real phenomena and teach real facts, but its fundamental intent is to provide a basis of skills and attitudes and a relation to nature, rather than facts as such. In a school system that relies heavily on memorized factual content this will be a delicate undertaking.

We have noted the prevalence of a dual view of nature or reality which was especially striking because the two views expressed seem to us contradictory, although accepted simultaneously by our subjects. If this paradox is new in Nepal, it is certainly not new to the West. The same ambivalence has run through Western thought at least since the early scientific revolution, and it is still with us. What scientist in the West has not heard the question, But how can you be a scientist and still accept that view? or has not at one time or another agreed to speak on "Science and Religion"?

Yet for the most part we in the West have been able to make our peace with

163

the complementary worlds of matter and spirit, of objects and values. Through careful delineation of boundaries, conscious and unconscious compartmentalization, or reinterpretation, and a variety of intellectual nonaggression pacts, a reasonably secure and peaceful coexistence has been achieved, so that this particular dualism no longer poses serious problems for the Western scientist or student. Can others be helped to achieve or preserve a coexistence that does not violate their cultural values as they try to assimilate our Western science and scientific viewpoint?

We propose, as one step, that science be presented as a "second culture," complementing that already present rather than replacing it, and taught in the spirit in which a second language is taught—to be learned and used, certainly, but not to the exclusion of the student's native tongue. This will require a very different orientation from that commonly found in most Asian schools, or indeed from that characteristic of most Asian-American relationships, even if it does not mean great changes in school curriculum. Beginning with the earliest missionary schools and continuing through the period of colonial schools, the attitude, and often the intent, of Western education has been that a "primitive" or "decadent" civilization is to be replaced with a more modern and "better" one (12). This attitude tends to continue even though colonialism is no longer a force behind it, and it tends to be particularly strong in science teaching, for science is taken to be the one really unique and powerful offering of the Western world. In fact, however, the purpose of education, whether in Nepal or elsewhere in Asia, is no longer to destroy one civilization, or even one set of ideas, in order to replace it totally with something that is conceived to be better; to proceed in that direction, or with that implicit attitude, is to create unnecessary difficulties

along the way. An implacable either-or approach, leading to a direct confrontation between traditional attitudes and a modern and very foreign approach to knowledge, invites conflict both within the student's own mind and between him and his elders in the community.

As has been seen too often, such a conflict results at best in a draw, which alienates from one world without really admitting into the other. We propose to avoid or postpone this confrontation by starting early science instruction with simple observations of ordinary things and events—observations which stimulate and use the child's latent curiosity, which anyone can make, and which demand no special or formal interpretation in cosmological or philosophical terms. Instead, this approach will provide a foundation of skills, of attitudes toward observation, and of specific observations upon which a more formal knowledge of science may later be built. In making this proposal we accept a complementarity of views as natural and perhaps as inevitable.

We are mindful of certain arguments that favor, in principle, the opposite alternative of immersing the student in the Western scientific culture, through study in the West, and demanding that he learn it and conform to it totally. Of course, this "total immersion" would not be possible for most Nepalese children or adults, the vast majority of whom do not in any case expect or wish to become scientists. Beyond this lies the fact that any who were to succeed in such a "total immersion" and then return to work and live in Nepal would be sure to find themselves seriously alienated. To a considerable degree this does happen to Asian graduate students who leave home to study in the United States and then, partly because of the consequent alienation, find themselves unenthusiastic about returning home. Once immersed, one is more comfortable to remain so.

We are mindful, also, of a seeming contradiction in our proposal. If extensive social and cultural changes are bound to accompany the introduction of science, is it wise to ignore this in preparing the child for learning science? We believe that it is. Of course, some kind of accommodation between the scientific revolution and Nepalese culture must and will eventually be reached if science is introduced at all. This is a complex matter which must evolve slowly within the Eastern cultures as it did in the West. Experience suggests that this accommodation will not be most easily achieved simply by substituting the one for the other, and particularly not during the school years when the children are immersed in the intellectual and physical environment of the village. It is important to them and to the village that they remain at peace there. Moreover, an eventual accommodation should be based upon real science well learned rather than on a set of memorized facts and formulas learned under stress.

To refer again to the analogy of language teaching, recent experience seems to show that the attitudes and techniques used in teaching English to, say, Urdu-speaking children, where no question of substitution or conflict arises, are the most effective in teaching a standard English to children, in Hawaii or Georgia, for example, who speak a "substandard" dialect of English. It is easier for these children to learn standard English when it is presented as a second language, not as a substitute for their own "incorrect" dialect, which of course they continue to need in their own community.

Elementary Science Instruction

A detailed program of instruction in science (many might prefer to call it pre-science) is under development in Nepal, partly as a result of the observations reported above. It would be premature at this time to anticipate its final dimensions or content, but we may say a few things about its methods and goals. Emphasis will be on an observational approach to phenomena which are familiar to anyone, or which can readily be produced by anyone, and a progression of skills and experiences will be built up, encompassing classification, measurement, generalization, inference, and quantification, and leading ultimately to the design and execution of elementary but conceptually more or less sophisticated experiments by the students. The material presented will contain some specific information intended for later use, but this will not receive major emphasis and need not be memorized. In addition to the observational material, some history of science will be introduced, in essentially anecdotal form, to show that knowledge of "books and old people" is really a record of observations and interpretations made by real people.

The educational system of Nepal, like that of many other countries, is fairly rigid and is not amenable to rapid change or experimentation. Nevertheless, we see it as necessary that improvements, to be lasting, be developed within the existing system, however time-consuming that may prove to be; hence, such efforts must have the understanding and active support of those persons who are in positions of leadership and authority within the system. The program we are proposing will be developed within the Education Ministry and tried out in the laboratory schools of the College of Education of Tribhuwan University. Further trial and development with the help of Peace Corps teachers, now working in many schools throughout the country, is anticipated. These Peace Corps volunteers are enthusiastic, well accepted, and devoted to the improvement of education. If the program has the endorsement of the government, they will be ready to try it out with very little feeling of hesitancy because the teaching methods are new and different. Moreover, there is reason to hope that the interest in better science teaching which has resulted from the American experimental curricula in secondary school science will result in changes and improvements in Asia, and perhaps in Nepal, affecting the teaching of science at the secondary level. These American courses, as well as other studies (*13*), emphasize observation and experiment. We hope that a program such as the one we propose for Nepal at the elementary levels will provide a useful preparation for more formal course changes patterned on the American model. In fact, we believe that some such preparation will be found necessary if science courses based on the American experimental courses are to maintain their spirit and emphasis as they are adapted for use in Asia.

In concluding, we must emphasize that much of what has been said is tentative, based as it is on a limited pilot study. Yet the study does indicate that research of this nature can provide needed perspective for the improvement of science teaching in non-Western countries. We hope it will lead to more study and discussion, with regard both to Nepal and to other developing countries.

References and Notes

1. R. Gruber, *Science and the New Nations* (Basic Books, New York, 1961).
2. U.S. Agency for International Development, *Science, Technology and Development: The U.S. Papers Prepared for the United Nations Conference on Applications of Science and Technology for the Benefit of the Developing Nations* (Government Printing Office, Washington, D.C., 1962–63).
3. F. E. Dart, *Foreign Affairs* **41**, 360 (1963).
4. B. Malinowski, in *Science, Religion and Reality*, J. Needham, Ed. (Macmillan, New York, 1925).
5. F. L. K. Hsu, *Religion, Science and Human Crises* (Routledge, London, 1952).
6. E. E. Evans-Pritchard, *Witchcraft among the Azande* (Oxford Univ. Press, London, 1937); J. Needham, *Science and Civilization in China* (Cambridge Univ. Press, London, 1954).
7. V. Barnouw, *Southwestern J. Anthropol.* **11**, 15 (1955).
8. See, for example, J. M. Roberts, M. J. Arth, R. R. Bush, *Amer. Anthropologist* **61**, 597 (1959).
9. Here we consider only children old enough to be capable of abstraction. See, for example, B. Inheldet and J. Piaget, *The Growth of Logical Thinking from Childhood to Adolescence* (Basic Books, New York, 1958).
10. Newars refer to four or five different kinds of thunder, of which this is one. Each has a name and is distinguished from others by the quality of sound perceived.
11. R. M. Gagné, *Science* **151**, 49 (1966); see also, "Science—a Process Aproach," *AAAS Comm. Sci. Educ. Pub.* (1965).
12. See, for example, R. I. Crane, in *The Transfer of Institutions*, W. B. Hamilton, Ed. (Duke Univ. Press, Durham, N.C., 1964).
13. "Guidelines for the Development of the Laboratory in Science Instruction," *Nat. Acad. Sci.–Nat Res. Council Publ. No. 1093* (1962).
14. We gratefully acknowledge the support and assistance of the Government of Nepal, the United States Education Foundation in Nepal, and the Center for Cultural and Technical Interchange between East and West in Honolulu in carrying out this study.

Chapter Four

MODELS, ASSORTED

Your physics text describes analog models, working models, and mathematical models, which are used by scientists in the formulation of theories. Each model gives only a partial or approximate representation of a phenomenon, but it also affords a significant insight. Therein lies the beauty of models: by stripping away some complexity, they lead to understanding. Often several complementary models, each inadequate of itself, can be combined into one more comprehensive picture, as in the wave-particle theory of light, or in Koestler's threefold analog model for the scientist (Chapter 1). Models are used outside physics, in all areas of human experience, even though that term may not be applied.

Analogies, the type of model illustrated first, are a commonly used tool in art and literature. One person or object evokes the image of another, perhaps bringing out a particular characteristic or highlighting a certain property. Occasionally these analogies are developed to the point where one may recognize an analog model that has analytical and predictive power. George Orwell's Animal Farm is a brilliant example, too long to be included here. Still better known is Jaques' speech comparing life to a theatrical performance in Shakespeare's "As You Like It." This model would lead you to predict that an individual has little control over his fate because the "lines" have all been written in advance.

Desmond Morris' picture of man in the "Introduction" to The Naked Ape strips away the complexity of the social roles stressed by Shakespeare. Read the book itself to find out how well this analog model fits. Undoubtedly, the name of the model conjures up certain images in your mind.

An application of the analog model concept to the social sciences is exploited by Robert E. Kuenne, who might ask, "Does the economy in its relation to society resemble a separable mechanism such as an automobile motor, or is it more like a vital organ, such as the heart of a man?" With today's broadening use of the social sciences in government, education, and business, it is becoming clear that preference for one analog model over the other can have far-reaching social consequences. Yet, at the present state of knowledge, this preference usually reflects the personal philosophy of individuals rather than a concensus based on scientific evidence. Have you encountered the conflict between these two approaches? Which model do you prefer?

Considerably less controversial are the four "fluids" described by Harold L. Davis. You may very well accept only blood as really being a fluid, but the report presents convincing evidence that a galaxy, the Earth's mantle, and traffic can also be treated profitably through the "fluid" analog model. Note that two of these examples represent a reversal of your text's "Many Interacting Particle Model": in the MIP model, a fluid is described as an aggregate of particles; here an aggregate of "particles" such as stars or cars is described as a fluid. You probably have had personal experiences with the traffic "fluid." Does the analog help you to understand traffic jams?

The next three selections in this chapter describe working models. These differ from the analog models in that the working model is made from postulated or idealized parts, rather than depending on a correspondence with parts of a real but different system. Dan O'Neill's cartoon presentation of a working model for hair growth, for instance, seeks to explain changing fashions in hair styles by means of the ideal "hollow head."

The Freudian model for character structure described by Clara Thompson explains human behavior through the three postulated entities of ego, id, and superego, which were invented for that purpose by Sigmund Freud. The very similar "adult, parent, child" model proposed by Eric Berne in Games People Play (Grove Press, New York, 1964) has proven itself widely popular in recent years.

Chester A. Lawson goes beyond Freud and seeks a physiological basis for human behavior. He has invented a working model for the functioning of the brain, which he divides into several parts with postulated properties. The parts interact through electric signals (called "traveling potentials" by Lawson) whose detection and identification may make it possible to confront this model with experimental data.

Mathematical models, which are so important in physics, have entered the social sciences only recently, because they depend on the availability of quantitative data. In the last article, David Gale describes how economists may go about constructing mathematical models and illustrates his thesis with a very simple example. Does this procedure fit into the mechanistic or organic alternatives defined by Kuenne earlier in this chapter?

About the Contributors

WILLIAM SHAKESPEARE (1564-1616) was the greatest dramatist writing in the English language. He frequently used analogies to convey his meaning effectively and picturesquely.

DESMOND J. MORRIS (1928-) is a zoologist, now turned author, who was for many years Curator of Mammals for the Zoological Society of London. He has written several books for the general public, including children's books. Dr. Morris' side interest is archaeology.

ROBERT E. KUENNE (1924-   ) is director of the General Economic Systems Project. The Project is producing a series of textbooks to further the study of economic models and their applicability to the analysis of real-world problems.

HAROLD L. DAVIS (1925-   ) is Editor of Physics Today, official publication of the American Institute of Physics. He was trained as experimental physicist and worked for a time on the design of nuclear reactors for aircraft propulsion before serving in editorial positions with several science-oriented magazines. Dr. Davis has side interests in mental health and cybernetics.

DAN O'NEILL (1942-   ) is a cartoonist in San Francisco. His cartoon strip "Odd Bodkins" was accepted by the San Francisco Chronicle syndicate only after it had been ignored by many newspaper editors. Yet, when the paper recently discontinued the strip, public clamor demanded and achieved its reinstatement.

CLARA M. THOMPSON (1893-1958) was a physician and psychiatrist in New York. She taught at several graduate institutions and was led, by questions from students, to consider the interaction of experiment and theory in the evolution of psychoanalysis as a science.

CHESTER A. LAWSON (1908-   ) is Director for Life Science of the Science Curriculum Improvement Study, University of California, Berkeley. His early work in genetics was soon replaced by an interest in general education at the college level, while he served as professor and department head in Natural Science at Michigan State University. Now his interests lie in elementary education, the evolution of ideas, and brain mechanisms.

DAVID GALE (1921-   ) is Professor of Mathematics, Economics and Industrial Engineering at the University of California, Berkeley. For some years previously he served as head of the mathematics department at Brown University. He specializes in mathematical economics and the theory of games.

"All the world's a stage...." From <u>As You Like It</u> by
William Shakespeare, Act II, Scene VII.

All the World's a Stage

by William Shakespeare

Jaques: All the world's a stage,
And all the men and women merely players.
They have their exits and their entrances;
And one man in his time plays many parts,
His acts being seven ages. At first the infant,
Mewling and puking in the nurse's arms.
Then the whining schoolboy, with his satchel
And shining morning face, creeping like snail
Unwillingly to school. And then the lover,
Sighing like furnace, with a woeful ballad
Made to his mistress' eyebrow. Then a soldier,
Full of strange oaths, and bearded like the pard,
Jealous in honour, sudden and quick in quarrel,
Seeking the bubble reputation
Even in the cannon's mouth. And then the justice,
In fair round belly with good capon lined,
With eyes severe and beard of formal cut,
Full of wise saws and modern instances;
And so he plays his part. The sixth age shifts
Into the lean and slippered pantaloon,
With spectacles on nose and pouch on side,
His youthful hose, well saved, a world too wide
For his shrunk shank; and his big manly voice,
Turning again toward childish treble, pipes
And whistles in his sound. Last scene of all,
That ends this strange eventful history,
Is second childishness and mere oblivion,
Sans teeth, sans eyes, sans taste, sans everything.

*

Introduction

There are one hundred and ninety-three living species of
monkeys and apes. One hundred and ninety-two of them are
covered with hair. The exception is a naked ape self-named
Homo sapiens. This unusual and highly successful species spends
a great deal of time examining his higher motives and an equal
amount of time studiously ignoring his fundamental ones. He
is proud that he has the biggest brain of all the primates, but
attempts to conceal the fact that he also has the biggest penis,
preferring to accord this honour falsely to the mighty gorilla.
He is an intensely vocal, acutely exploratory, over-crowded
ape, and it is high time we examined his basic behaviour.

I am a zoologist and the naked ape is an animal. He is therefore fair game for my pen and I refuse to avoid him any longer
simply because some of his behaviour patterns are rather complex and impressive. My excuse is that, in becoming so
erudite, *Homo sapiens* has remained a naked ape nevertheless;
in acquiring lofty new motives, he has lost none of the earthy
old ones. This is frequently a cause of some embarrassment to
him, but his old impulses have been with him for millions of
years, his new ones only a few thousand at the most—and
there is no hope of quickly shrugging off the accumulated
genetic legacy of his whole evolutionary past. He would be a
far less worried and more fulfilled animal if only he would
face up to this fact. Perhaps this is where the zoologist can
help.

One of the strangest features of previous studies of naked-

9

ape behaviour is that they have nearly always avoided the obvious. The earlier anthropologists rushed off to all kinds of unlikely corners of the world in order to unravel the basic truth about our nature, scattering to remote cultural backwaters so atypical and unsuccessful that they are nearly extinct. They then returned with startling facts about the bizarre mating customs, strange kinship systems, or weird ritual procedures of these tribes, and used this material as though it were of central importance to the behaviour of our species as a whole. The work done by these investigators was, of course, extremely interesting and most valuable in showing us what can happen when a group of naked apes becomes side-tracked into a cultural blind alley. It revealed just how far from the normal our behaviour patterns can stray without a complete social collapse. What it did not tell us was anything about the typical behaviour of typical naked apes. This can only be done by examining the common behaviour patterns that are shared by all the ordinary, successful members of the major cultures —the mainstream specimens who together represent the vast majority. Biologically, this is the only sound approach. Against this, the old-style anthropologist would have argued that his technologically simple tribal groups are nearer the heart of the matter than the members of advanced civilizations. I submit that this is not so. The simple tribal groups that are living today are not primitive, they are stultified. Truly primitive tribes have not existed for thousands of years. The naked ape is essentially an exploratory species and any society that has failed to advance has in some sense failed, 'gone wrong'. Something has happened to it to hold it back, something that is working against the natural tendencies of the species to explore and investigate the world around it. The characteristics that the earlier anthropologists studied in these tribes may well be the very features that have interfered with the progress of the groups concerned. It is therefore dangerous to use this information as the basis for any general scheme of our behaviour as a species.

Psychiatrists and psycho-analysts, by contrast, have stayed nearer home and have concentrated on clinical studies of mainstream specimens. Much of their earlier material, although not suffering from the weakness of the anthropological information, also has an unfortunate bias. The individuals on which they have based their pronouncements are, despite their mainstream background, inevitably aberrant or failed specimens in some respect. If they were healthy, successful and therefore typical individuals, they would not have had to seek psychiatric aid and would not have contributed to the psychiatrists' store of information. Again, I do not wish to belittle the value of this research. It has given us an immensely important insight into the way in which our behaviour patterns can break down. I simply feel that in attempting to discuss the fundamental biological nature of our species as a whole, it is unwise to place too great an emphasis on the earlier anthropological and psychiatric findings.

(I should add that the situation in anthropology and psychiatry is changing rapidly. Many modern research workers in these fields are recognizing the limitations of the earlier investigations and are turning more and more to studies of typical, healthy individuals. As one investigator expressed it recently: 'We have put the cart before the horse. We have tackled the abnormals and we are only now beginning, a little late in the day, to concentrate on the normals.')

The approach I propose to use in this book draws its material from three main sources: (1) the information about our past as unearthed by palaeontologists and based on the fossil and other remains of our ancient ancestors; (2) the information available from the animal behaviour studies of the comparative ethologists, based on detailed observations of a wide range of animal species, especially our closest living relatives, the monkeys and apes; and (3) the information that can be assembled by simple, direct observation of the most basic and widely shared behaviour patterns of the successful mainstream specimens from the major contemporary cultures of the naked ape itself.

Because of the size of the task, it will be necessary to over-simplify in some manner. The way I shall do this is largely to ignore the detailed ramifications of technology and verbaliza-tion, and concentrate instead on those aspects of our lives that have obvious counterparts in other species: such activities as feeding, grooming, sleeping, fighting, mating and care of the young. When faced with these fundamental problems, how does the naked ape react? How do his reactions compare with those of other monkeys and apes? In which particular respect is he unique, and how do his oddities relate to his special evolutionary story?

In dealing with these problems I realize that I shall run the risk of offending a number of people. There are some who will prefer not to contemplate their animal selves. They may consider that I have degraded our species by discussing it in crude animal terms. I can only assure them that this is not my intention. There are others who will resent any zoological invasion of their specialist arena. But I believe that this approach can be of great value and that, whatever its short-comings, it will throw new (and in some ways unexpected) light on the complex nature of our extraordinary species.

THE ECONOMY: MECHANISM OR ORGANISM?

The first decision that we must make after deciding to map the economy is the type of model we will produce, and in the initial stage of our analysis this depends upon our view of the economy. Does it resemble a *mechanical* entity, placed within the whole social complex but removable from it for purposes of analysis; with a stable structure that can be changed within broad limits by its designers; which functions under the regime of inner laws built once-for-all into that structure; and which is essentially insulated from external impacts except those it is designed to accept as inputs? Or can it be viewed only as one *organ* of the larger social organism, related holistically to that organism and incapable of worthwhile physiological analysis removed from it; subject to disturbance from many forms of external influences and possessing some potential to adapt its structure to them inasmuch as its inner processes are life processes that do not freeze its structure for all time, but give it the capability of adaptation to external impacts? Further, does it possess within itself an inner law of change through time, such as growth and decay, or some more complicated *process* involved with the mere passage of time?

Let us seek to clarify these distinctions by broadening our outlook a bit. We may define a *society* as a family of *systems*. These systems may be divided for convenience into three types:

1. the *culture system*, or the whole complex of values, beliefs, symbols, goals, traditions, and ways of looking at the world that the society embraces;

2. *social systems*, or those systems in which two or more persons interact meaningfully for the accomplishment of individual and social ends, and in which the individual acquires *role-expectations* for his own and others' conduct; and

3. *personality systems*, most particularly those characteristics of the individuals in the society that interact with the culture and social systems to determine the *actions* of the individual. Included in this category are all kinds of activities (thought, for example), which have some social (as opposed to purely individual) meaning.

These three sets of systems are intimately interconnected, mutually affecting and being affected by all other systems. The economy, for example, is a member of the set of social systems, and contains within itself other social systems such as corporations, labor unions, markets, and the like. These component systems interact with other social systems such as governments, contain individuals who have been mightily affected by the culture system within which they move, and enforce upon these individuals certain prescribed manners of behavior toward others. But, if we found in our initial view of the economy that its detail was overwhelming and that any meaningful analysis required greater simplification, our difficulties would be compounded a millionfold were we to attempt to analyze as one entity in full detail the largest system—society.

Our decision, therefore, to remove one social mechanism—the *economy*—from the matrix of other systems in order to study it, was a step toward simplification even prior to our decision to lessen economic detail. The word *remove* implies that many of the lines of mutual influence must be made paths of one-way causation, along which feedbacks from the affected social system to the economy are not permitted. Now, if the economy, and more broadly, the society, can be treated in the light of a *mechanism*, this removal of the economy and the necessary distortions it requires are less punitive

177

analytically, for the frozen structure of the market mechanism does not change as we lift it out and cut its affiliations with many other systems. We have merely blocked off from it a variety of inputs, which it formerly received and which may have been processed by it, but to which its inner structure did not respond nor its in-built rules of operation alter in response. That is, if our "vision" of the economy is Newtonian in inspiration, reflecting a view of the economy as in the nature of a solar system, following stable laws of motion essentially unaffected by the laws of other solar systems, and with a structure which, if changing through time, is so stable that the changes can be ignored for long time periods, we may regard our operation as non-distorting.

A great temptation exists to adopt this view of the market economy for a simple but compelling reason: there exists a body of logical and mathematical techniques, developed by physicists, engineers, astronomers, and mathematicians in their analyses of physical systems, which is ready to hand for the physiological analysis of an economic mechanism. We can bring economic analysis under the regime of forces striving for an equilibrium, study changes in the equilibrium when inputs are changed to get insights into the assumed mechanism, and observe the performance of the mechanism as it functions through time, by applying mechanical statics and dynamics to its analysis. This approach pushes our analysis into a search for the principles of converting inputs to outputs that inhere in the structure of the mechanism, and therefore into an analysis of the rigid construction of the machine. The physiology of the system is, for purposes of such analysis, wholly self-contained (given inputs from the outside), and very closely related to the anatomy of the system. The outlook diverts our analytical attention from features of performance of the economic system that are *structural* responses to outside stimuli or changes inspired by inner laws of development within the system itself. That is, it ignores most *organic* aspects of the system.

If we consider the economy to be analogous to an organ in the organism that is society, the removal of the organ by definition is a dissection of the social body and automatically the organ ceases to function, or, at least, to function in its normal manner when all its linkages to other systems are intact. If we cut the ties of the economic system to the legal system, and in so doing, fail to take account of such entities as the body of antitrust law, the functioning of firms and markets may be quite different from what actually occurs in the American economy. Removing linkages with other social systems has the same effect as this abstraction from the legal system. If we cut the ties of the economy to the culture, which contains, for example, a set of attitudes toward work which we label the *Puritan work ethic*, any

178

study of labor supply in the economy may be misleading. Further, any vision of the economic system that does not allow structural change in a living process—for example, the changes in attitudes to work, that arise as a labor force gains experience and more income, or the responses of price to changes in demand as oligopolists in a market become more accustomed to the reactions of their rivals in a learning process that can only be characterized as organic changes in the structure of the market—may yield results that are fundamentally deficient.

Thus, we may discern two basic and antagonistic *visions* of the nature of the economic system. The first, which we may call the *mechanistic* view, has been in the ascendant among professional economists at least since the time of David Ricardo, although there were some strong strains of the second, or *organic*, view in the work of such leading theorists as Alfred Marshall. The organic view has been best expressed by *historicist-institutionalist* analysts like Thorstein Veblen, John R. Commons, Wesley Clair Mitchell, Cliffe Leslie, and the German historical school of economists. But its failure to become dominant among economists may be ascribed to a number of factors.

First, and perhaps most importantly, the adherents of the organic view have not succeeded in devising a body of analytical methods with the degree of rigor and with the success in the derivation of fruitful theorems about reality that the mechanistic school has done. Indeed, many of the organic view's adherents have attempted to show that it is impossible to develop such a body of formal methods, so that practically speaking only intensive *ad hoc* studies of historical process can derive insights into the functioning of the economy, the results of which possess only a limited capability for generalization given the organic changes in the system constantly occurring. The approach to a study of history without a body of formal techniques admittedly requires a more subtle mind, a much broader spectrum of detailed knowledge of the history of society, and promises fewer concrete and purportedly generalizable results, than the adaptation of mechanical statics and dynamics to social phenomena. In an age of "science" and logical positivism the appeal of the mechanical view to the body of professional economic analysts has been much more powerful.

Second, to an important degree the clash between the mechanistic and organic views occurs because of a difference in viewpoint about the *problems* that should be attacked by the economics profession. The adherent of the mechanistic viewpoint tends to focus upon shorter-run questions, such as how prices are formed by the mechanism under various constraints concerning adjustments, how quantities are arrived at in periods that are quite short historically, decisionmaking for the consumer and firm over short periods, and so forth. For such periods even the adherent of the organic approach might agree that the economic system does not change fundamentally under its own organic laws of change, nor does it have time to react greatly to external impacts from the culture or other systems. For short periods, therefore, speaking historically, the social organ may be viewed adequately as a mechanism. The historicist-institutional school argued that these are not the problems in which the field of economics should be interested, and therefore stressed the need to treat the economy organically. There was, then, a good deal of *problem relativity* in the conflict, to the extent that many economists would no doubt agree that short-run problems are best attacked by the mechanistic model, whereas historically long-run problems are best attacked by organic models.

Thirdly, there is an important difference in implications for policy in the two viewpoints, which has biassed professional economists toward the mechanistic methods. Analytical treatment of the economy as a mechanism develops in the analyst an attitude of *social engineering* concerning normative approaches to his subject. If his approach to analysis is that of dealing with a mechanism, his attitudes toward social problems arising from its functioning may well be that they seem capable of being handled by proper alterations in the structure of the mechanism. The abstractions of the mechanism from its ties with other social systems tend to be ignored or at least subordinated, so that the analyst with understandable desires to effect policy changes is consciously or unconsciously drawn to the mechanistic framework of analysis.

On the other hand, those who embrace the organic viewpoint tend to be more cautious in their approach to policy recommendations. At one extreme they view man as caught up in a web of cultural and other institutions that impede the rational adjustment of society to its externally imposed needs and make futile any social engineering attempts to improve his lot in the short run. It is ironic that although institutionalism was in revolt against the social Darwinism rampant at the end of the century, and although it was mightily affected by reform movements like Progressivism and Populism and by such policy-oriented intellectual movements as philosophical pragmatism, John Dewey's instrumentalism, and Roscoe Pound's sociological jurisprudence, the fundamental organic outlook which underlay it prevented it from becoming a strong proponent of legislated or educated reform.

The mechanics of four fluids

Harold L. Davis • *Senior Editor*

A highlight of the spring meeting of the American Physical Society in Washington was a well-attended session on unusual applications of fluid mechanics.

Four fluids were covered: galactic matter, blood, the earth's mantle, and automobile traffic. This classical Newtonian branch of physics, it is clear, is still very much alive.

1. HEAVENLY BODY PLASMA

The most ambitious investigation imaginable in fluid mechanics—in terms of physical dimension—is being carried out by C. C. Lin of MIT.

With outer space as his lab, in effect, Lin is seeking to explain the familiar spiral structure of galaxies in terms of a wave phenomenon involving the collective interactions of the 100 billion stars that make up a typical galaxy. He looks on the stars in a galaxy as particles constituting a moderately dense plasma. (The ratio of particle volume to empty space in a galaxy is about the same as in a plasma containing 10^{15} protons per cm^3 in a laboratory plasma-physics experiment. The important difference is that the particles of the galactic plasma interact via gravitational forces, rather than through electromagnetic forces.)

Over the last decade astronomers have been divided by two rival explanations of the magnificent spiral structure observed in many galaxies. The original views, proposed by Swedish astronomer Bertil Lindblad—and supported by a mass of impressive evidence from Lin—is that the spirals are density waves wheeling about the galactic center through a homogeneous volume of stellar plasma. The second view is this: the spirals are material in nature—they represent permanent concentrations of stellar material that rotate about the galactic center. Astronomers have speculated that the galactic magnetic field (about 5 microgauss) might give rise to the observed spirals by means of hydromagnetic containment effects.

But a glaring difficulty of the material spiral theory has been its disagreement with the fact that the rotation of galaxies is differential—stars near the center and the edge move with roughly the same circumferential velocity, so that the inner stars have a much faster rate of rotation than the outer ones. If the spirals are material structures, it is impossible to understand how they could maintain their shapes, given the differential rotation, for more than one or two revolutions.

By contrast, Lin and others have succeeded in developing a quantitative model of the density-wave theory that is consistent with the differential rotation of the galaxy and leads to detailed observational predictions, some of which have recently been confirmed.

In their opinion, stellar matter or plasma is uniformly distributed throughout the galaxy. Spiral density waves move through this plasma, causing peak fluctuations of about 5 percent in density. At the wavecrest, this increase in density is enough to trigger gas clouds on the verge of gravitational collapse to give birth to new stars. Each of these stars is ten thousand times brighter than a typical older star, such as our sun. These populations of young stars, Lin argues, conveniently outline the crests of the spiral-wave pattern viewed in our telescopes.

The waves make a complete trip around the galaxy once every 450 million years, which is actually slower than the rotational rates of stars in most parts of the galaxy. Stars in the vicinity of our sun (two thirds of the maximum radius) have a galactic year of 200 million years.

2. CARDIAC COCKTAIL

The problem of unwanted clotting in the bloodstream intrigues aerodynamicist Arthur Kantrowitz because it is velocity dependent—that is, clotting occurs only in blood flowing slower than certain critical speeds.

Kantrowitz (who is director of the Avco Corp. Everett Research Laboratories) observed that the clots that appear on prosthetic material placed in the bloodstream (such as sutures or artificial heart valves) probably form most copiously in regions of stagnation or reverse flow. He suggests using the well-established science of fluid mechanics to make sure that future prosthetic devices have optimum hydrodynamic properties.

From the clinical viewpoint the problem is far from trivial. Any artificial substance introduced into the bloodstream is quickly covered with a lining of clotted blood. Eventually, tissue forms across the surface and incorporates the clot material.

The danger is that before this happens pieces of clot may break away and be carried to the heart or the brain, causing thrombosis. Artificial heart valves, for instance, are designed without regard for hydrodynamics and, as a result, excessive clotting is believed to occur at stagnation points—leading to the formation of long streamers of clots that can easily break off.

Statistics indicate that 30 percent of all patients with artificial valves have trouble with clots breaking off. In one third of these cases the trouble proves fatal. The problem is expected to become even greater when bigger and more complex prosthetic devices, such as artificial pumps, come into common use.

Although a few physicians have published reports on the mechanism of clotting, their experiments, Kantrowitz observed, have been wanting in sophistication and have failed to correctly simulate the blood chem-

istry and provide a meaningful model of the fluid mechanics.

These objections were overcome in experiments recently begun at Avco in which the distribution of clotting is observed on a flat plate when the bloodstream from the neck artery of a dog is directed perpendicularly to the plate. For this configuration, in which the hydrodynamic flow can be precisely calculated, the Avco group has sought to explain why clotting is flow dependent by testing three hypothetical mechanisms:

1. A minimum reaction time determined by a chain of chemical events.

2. Diffusion of the clotting substances to the foreign surface.

3. Velocity-dependent shearing of molecular bonds that hold the clot material to the surface.

Kantrowitz's findings indicate that the last two mechanisms are involved but the first is not. Microphotography shows that the clots build up from the deposit of blood platelets that diffuse to the surface and adhere to it with a force typically as small as a single molecular bond. What causes the platelets to deposit out on foreign surfaces and not on the natural membranes that line blood vessels is not yet known.

3. WORLD ON THE ROCKS

The floor that holds up both the continents and the oceans—the upper mantle of the earth—has long been known to consist of high-density basaltic rock. A variety of evidence reported at the meeting by Donald L. Turcotte of Cornell University is now compelling geophysicists to consider the possibility that this massive layer of rock, extending down 500 km, behaves as a viscous fluid in which huge convective currents circulate under the driving force of the earth's thermal gradient.

The earth's temperature increases with depth and, because of radioactive heat release, this gradient is thought to be especially steep (as much as 20° C/km) in the outermost regions that include the upper mantle. Since the rock of higher temperature in the depths of the mantle expands to a lower density than the colder rock nearer the surface, the former experiences a buoyant force that drives it up through the colder material.

Like the convection currents that arise in the air in a room heated near the floor, these convection currents are thought to extend vertically through the depth of the earth's mantle and horizontally for thousands of miles under the earth's crust. The existence of such a circulatory flow in the earth's mantle provides a neat explanation of some puzzling geological features, for example, the oceanic ridge structure that runs the length of the major oceans and is associated with volcanic and earthquake activity.

Paleomagnetic studies of samples of the surface in the vicinity of the ridges have recently established that the ocean floor recedes in both directions from the center of a typical ridge at the rate of 1-4 cm/year. Turcotte and others argue that the oceanic ridges (and perhaps some of the younger mountain ranges) mark the loci of the rising hot legs of the subterranean currents.

Along these ridge lines, hot, near-molten rock continually forces its

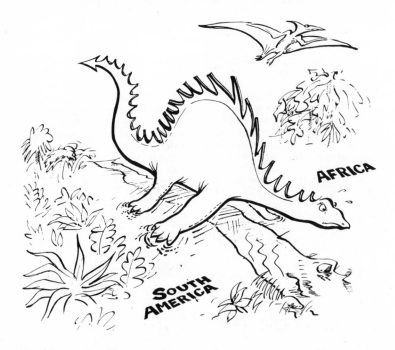

way up, occasionally causing lava to erupt through the cold crust and inexorably forcing the floor of the sea to move away from the ridge in either direction. The motion of the floor forms the topmost layer of the convective flow.

The descending, cold legs of the loops are believed to be found at the deep ocean trenches; there is evidence that these 8,000-meter-deep canyons serve as sinks that return material to the lower depths of the mantle.

Quite often the trenches occur near the edge of the continents and the picture of the earth's mantle as a viscous fluid, divided up into about ten giant convective regions, has of late inspired new interest in the continental drift hypothesis. That idea, originally prompted by the geometrical correlation between the facing boundaries of, say, Africa and South America, has also received support from the excellent matching of paleomagnetic measurements found for pairs of corresponding locations in different continents.

It has been generally accepted that the continents, composed of lighter rock, "float" on the higher density upper mantle. What has been missing is a mechanism that could supply the energy needed to push South America 3,000 miles from Africa; convection currents, functioning like mammoth conveyor belts, seem to provide it.

Thus, it is possible that 200 million years ago North and South America, Europe, Africa and Greenland formed a single landmass that was split apart, the pieces being pushed away in various directions by the mid-Atlantic ridge.

Thermally driven flow loops in the mantle were first proposed in 1931 and were recently given a more rigorous form by Turcotte and others. The new models give calculated values of the heat flux from the ocean ridges that agree well with observations, assuming a fluid viscosity of 10^{21} poise for the rock of the upper mantle.

Still a matter of disagreement is the exact nature of the fluidlike behavior of the mantle. Turcotte assumes the mechanism is diffusion creep, which gives a Newtonian viscosity. Other researchers think plastic flow rather than fluid flow may be involved.

4. EXPRESSWAY BREW

A fundamental approach to predicting the behavior of traffic was recommended by Robert Herman of General Motors: consider a traffic stream to be a continuous fluid. The vehicle-driver units that make up the collective fluid, however, are controlled by psychological interactions and not by the usual laws of physics.

Thus, in one of the simplest models the velocity V, of a car is related to the velocity V_0 of the car preceding it by $V_1 (t+\tau) = CV_0 (t)$ where τ is a physical-reaction time delay and C is a coefficient in which the driver's personal habits and preferences regarding such factors as tailgating, braking rates, and back-seat advice are embedded.

These considerations lead to a mathematical description that resembles the Boltzmann equation used in the kinetic theory of gases. Such a model, Herman feels, can be helpful in understanding how a traffic jam occurs. The theory predicts that at some critical concentration (about 50 cars per mile) there will be a transition from individual flow to collective flow.

Up to this critical point, the flow is governed largely by the desires of the individual drivers. Beyond this point, traffic moves in a rather gelled state, in which the actions of each driver are determined by the properties of the fluid. As any rush-hour driver can confirm, the average speed of the traffic stream is independent of the wishes of the drivers and depends almost entirely on the traffic concentration and the probability of passing.

Such collective flow is characterized by instabilities and large fluctuations that are experienced as local traffic jams. The more serious instabilities, such as chain collisions, apparently can be triggered by the injection of relatively small perturbations in a traffic stream whose drivers have C coefficients neighboring on the average. Computer calculations show that the abrupt deceleration of a lead car from 60 mph to say 50 mph causes a deceleration wave (as measured by the distances between cars) to extend back along the traffic stream. The wave gains in amplitude as it travels until the distances between cars becomes negative.

Most people, Herman observed, habitually drive close to their limit of stability.

Although traffic problems go back to the time of the Caesars (there was talk of prohibiting chariots in the center of ancient Rome), the fact that the problem has grown to the point where the automobile is a major cause of death should inspire a large-scale basic research effort in this area, Herman feels. Not only do automobile accidents cause the greatest number of deaths in absolute numbers among teenagers and young adults, he says, but among older adults the risk of death per hour of exposure in an automobile is actually greater than for death by heart disease (in the sense that one is "exposed" to one's heart 24 hours a day).

Herman complained that despite the increased funds available for transportation problems little has been done to encourage fundamental studies of traffic problems. All too often, he contends, traffic engineers are shockingly unsophisticated in physical theory or mathematics. As a simple example, the duration of the yellow caution light is often so short that a car travelling within the speed limit at a certain distance from the intersection when the yellow flashes on is unable either to stop or to cross the intersection before the light turns red.

Herman cautioned that if the large-scale appropriations now being made for planning transportation systems are to be well spent, young graduates with good physics and mathematics backgrounds must be persuaded of the the possibilities for basic research in this field.

THE EGO AND CHARACTER STRUCTURE

Wᴵᵀᴴ ᴛʜᴇ ᴅɪsᴄᴏᴠᴇʀʏ ᴏғ ᴛʜᴇ ʀᴏʟᴇ ᴏғ ᴀɢɢʀᴇs sion and the study of repetitive patterns of behavior, the function of the ego eventually became the topic of study. In 1910 Adler made the first move in the direction of stressing the importance of the ego and its functions. Freud, as we have seen, had earlier assumed that ego drives were not subject to repression and therefore played no part in neurosis. He repudiated Adler's work on the basis that in stressing ego traits he was denying the importance of the unconscious and was, therefore, no longer dealing with the material of psychoanalysis.[1] Freud at the time certainly failed to see a fact which became clear to him in the 1920's, that ego drives might also be unconscious. So Adler's discovery made no impression on the main stream of psychoanalysis.

However, soon after 1910, Freud began to give some consideration to the ego. In 1911 he wrote a very important paper on "Two Principles of Mental Functioning" in which,

[1] For Freud at this time unconscious material was always related to libidinal drives.

59

while stressing again his already well established theory that the libido functioned according to the pleasure principle, he pointed out that ego drives seemed more under the influence of the reality principle. Out of the necessity of comprehending reality developed man's ability to observe, remember and think. In short, out of the need to cope with reality grew the ego as we know it. Ferenczi,[1] in 1913, elaborated the theme in his paper, "Stages in the Development of the Sense of Reality." He showed that in the process of growth, not only the libido went through stages; the ego also came only gradually to full comprehension of its function of reality testing.

He suggested that the ego emerges gradually, going through four preliminary stages before clearly becoming differentiated as an entity. The first stage is the period of "unconditional omnipotence." This is the situation of the child before birth when all wishes are gratified. Immediately after birth is the period of "magic hallucinatory omnipotence." At this time, he thought, the infant must feel that he only needs to wish something and it is there. As his needs become more complex, there are times of disappointment, but then he discovers that by cries and gestures he can produce results. This is the period of "omnipotence by magic gestures." And finally comes the stage of power through "magic thoughts and words." Gradually the feeling of omnipotence fades, although Ferenczi suggested that some people never give up the idea of magic and keep the illusion of power in the idea of free will.

At any rate, the ego and its function were beginning to be

[1] Sandor Ferenczi, *Sex in Psychoanalysis*, Ch. VIII.

noted. As has already been mentioned, the discovery of the importance of repressed aggression further focused attention on the ego. In the early 1920's Freud finally formulated a theory of the total personality, and the ego with its function of reality testing became the Ego of the Ego, Superego and Id.

Freud saw the newborn infant as chiefly Id, that is, masses of impulses without an organizing or directing consciousness. Contact with the world gradually modifies a portion of this Id and a small area of consciousness, the Ego, slowly emerges from it. It is developed out of the necessity for reality testing. However, the Ego is not synonymous with consciousness. Only a small part even of the Ego is conscious at any one time. A great part of the Ego as now defined exists outside of awareness but can readily be called into awareness when needed. This part was called the preconscious. Still another part of the Ego is unconscious and cannot readily be made conscious. This consists of the experiences and feelings which have been repressed. These experiences, by the fact of their repression, are somehow brought into more intimate contact with the forces of the Id.

The Id is a mass of seething excitement which cannot become conscious directly. Many of its forces never reach awareness, but from time to time portions of its energy can find some expression in the Ego by becoming connected with the memory-traces of repressed experience and thus participating in the formation of symptoms; becoming distorted as in dream symbols; or by undergoing modification chiefly as a result of the influence of the Superego as in sublimation. The Id Freud conceived of as of tremendous size in comparison with the Ego. He thought of it as the generator of energy,

191

the dynamo of the personality. It is somehow closely associated with the organic processes of the body.

In the course of time, the Ego takes over certain standards from the culture, chiefly through the influence of training by parents in early childhood. These standards become incorporated as parent images within the Ego as a part of itself and this part is called the Superego. It exercises a criticizing and censoring power. The functions of dream censor and resistance described in Freud's earlier writings are now seen as part of the Superego. The Superego, in brief, represents the incorporated standards of society. It includes the parental attitudes, especially as these attitudes were understood and interpreted by the child in his early years. It includes also the person's own ideals for himself, and Freud even indicates that certain phylogenetic experiences such as those described by Jung under the concept of the collective unconscious may also be part of the Superego. Much of the Superego is unconscious because it was incorporated by the child very early and without his awareness. This means that like all unconscious material this portion is not available for reality testing. This partially accounts for the irrational harshness of some of the attitudes of a man's conscience towards his behavior. For example, it may make him feel guilty about an act for which he consciously has no feelings of regret. Freud further attributes some of the harshness of the Superego to a theoretical relation to the death instinct. This is too complicated to be discussed here.[1] The Superego is an important construction. It is, in effect, Freud's way of talking about in-

[1] Sigmund Freud, *Civilization and Its Discontents*, Jonathan Cape & Harrison Smith, New York, 1930, Ch. 7.

terpersonal relations, and the influence of the culture on man's behavior.

The Ego, as Freud saw it, holds an executive position. Its function is to reconcile the Id, the Superego and the outside world. It must permit the Id to let off enough steam so that its forces are not a dangerous threat and yet not offend the Superego or run afoul of the outside world. It is, so to speak, the master of compromise. With the aid of the Superego, it makes the forces of the Id harmless by forming sublimations or reaction formations from them.

Character structure, as Freud saw it, is the result of sublimation or reaction formation. That is, it is formed unconsciously through the efforts of the Superego to bind the forces of the Id in such a way that the Ego accepts them, and its relation to the outside world is not jeopardized. It is, in effect, a defensive mechanism. Although the result, sublimation, seems to be a positive attitude of the Ego, it is formed primarily as a defense against instincts. Freud's philosophy of character makes it the result of the transformation of instinctual drives.

The concept of the individual described in *The Ego and the Id* is, as Freud himself says, merely a theoretical construction which may or may not have validity, but it had a far reaching practical result. In this work (*The Ego and the Id*) Freud for the first time shifted his interest from the libido to the activities of the Ego. This shift influenced the technique of therapy. Within a few years, analysts were to become less concerned with what happens to the libido and very much concerned with the ways in which the Ego defends itself.

Introduction

Thus, perception, represented as a pattern of traveling potentials, is a function of a sensory input pattern as well as of the way the nervous system operates. As the sensory input pattern changes one could expect the pattern of traveling potentials to change also.

Now let us assume that a pattern of traveling potentials can also occur in the motor cortex and that this produces impulses in the motor nerves, activating the effectors to produce a pattern of behavior. For this behavior to fit the environment, the pattern of traveling potentials in the sensory cortex must be integrated with the pattern of traveling potentials in the motor cortex. Thus we can conceive of brain function as a dynamic system consisting of patterns of traveling potentials in the sensory cortex areas maintained by impulses from the receptors. These patterns then produce and maintain patterns of traveling potentials in the motor cortex that order behavior. As the sensory input changes, the patterns in the sensory and motor cortex change, with the result that the behavior changes.

The behaving organism lives in a dynamic environment consisting of other behaving objects, and the organism must coordinate his behavior with that of others in order to survive within the environment. Sensory input gives the organism information concerning the state of the environment and this information must be translated into the appropriate behavior. Thus, somewhere between the pattern of traveling potentials maintained by sensory input, and the pattern of traveling potentials within the motor cortex, an integrating mechanism must operate that selects a particular motor cortex pattern on the basis of the sensory cortex pattern.

However, something more than an integrating mechanism is necessary, at least for man. Given only an integrating mechanism that matches sensory input with motor output, the organism's behavior would fit the environment, but the organism would be nothing more than a robot. A man does more than respond to particular stimulus with a particular behavior. He is oriented toward the future. He has goals and expectations which may select and control behavior. Furthermore, he learns to perceive, he learns new behavior patterns, and he learns to modify and correct existing behavior patterns. What sort of mechanism or process can accomplish all this?

The explanation I wish to propose assumes there must be patterns of traveling potentials in sensory and motor cortex areas, and that there is another cortex area that aids in integration. It also assumes the existence of a chemical memory record or code within the brain stem. This code, I suggest, consists of two functionally separate parts, one related to sensory input, the other related to motor output. Both sensory and motor parts of the code are presumed to be linear structures composed of elements linked together like beads on a string. Since the code is presumably contained in neurons in the brain stem, the elements of the code may be single neurons. If that is so, then the string of beads would be a string of neurons connected linearly. Whether or not that is the case is immaterial for the present. The important idea at this stage is the functional relation of the code elements, shown in Figure 2.

In Figure 2 the sensory cortex area is shown firing a pattern of traveling potentials in response to input from sensory receptors. This pattern is transmitted from the sensory cortex area to the brain stem, where it activates an element of the sensory code designated S. The activated S element of the code then does two things. It sends impulses to the integrating cortex area, and it activates a connected code element designated T. Code T then also sends a pattern of nerve impulses to the integrating cortex area. The portion of the activity in the integrating cortex area that results from Code S represents what the individual perceives. The portion that results from Code T represents what the individual expects to perceive next. Thus present and future are connected within the activity of the integrating cortex area. But one more element is needed — input from proprioceptors also contributes a

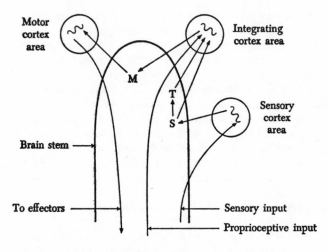

Figure 2

Schematic diagram showing the relation of the code in the brain stem to sensory and motor cortex areas. For fuller explanation, see the text.

portion of the pattern. This three-part system then transmits a pattern of impulses to the motor code within the brain stem. The portion of the motor code activated depends on the interrelated activities of the integrating cortex brought about by the S and T codes and the proprioceptive input. Activating the motor code fires a pattern of impulses in the motor cortex which then produces behavior. The behavior changes the relation of the organism to the environment, with the result that new sensory input reaches the sensory cortex. The revised pattern in the sensory cortex is transmitted to the brain stem, where it in turn activates the sensory code S. If the code was correct, this relayed input should now activate the T element of the code, that had previously been activated by the S element. The consequent activity in the integrating cortex area would constitute confirmation of the expectation, or reinforcement, and at the same time activate the next portion of the linear code to produce another expectation, and so on, through the whole coded behavior pattern.

Explaining the integration of human behavior and learning by this hypothesis involves a number of assumptions. These will be introduced as they are needed. The first is that the human infant inherits a general code that integrates his initial behavior and which may be modified and expanded by learning.

This assumption makes it possible for us to explain the eye-hand coordination shown by Gesell's infant subject, described earlier. The infant looked at, reached for, and grasped a cube placed on a table before him. He brought the cube to his mouth, moved it away again, and passed it from hand to hand. Looking at the cube presumably produces a pattern of impulses in the visual cortex, activating code element S, which fires a pattern of impulses in the integrating cortex area. Code S also activates Code T, which in turn fires a pattern of impulses in the integrating cortex area. The integrating cortex activity set off by the S pattern represents perception of the cube, which we have assumed is a primitive figure-ground unity for the infant. Code T represents an expectation, presumably the expectation of a tactile sensation upon grasping the cube.

From the moment the infant first sees the cube until he grasps it, the S and T patterns in the integrating cortex area presumably remain constant. However, from the beginning of reaching until the hand grasps the cube, there presumably is a continuous change in the proprioceptive input and thus a continuous change in the firing pattern of the integrating cortex area. Behavior is a function of the activity in the motor cortex and of the relation among the S, T, and proprioceptive portions of the activity in the integrating cortex area. Hence, the changing proprioceptive input causes and controls continuous changes in the behavior, and we see the infant reach for and grasp the cube.

Reprinted from AMERICAN SCIENTIST, Vol. 44, No. 1, January 1956

MATHEMATICS AND ECONOMIC MODELS

By DAVID GALE

Brown University

IF one looks back at the historical developments of mathematics and economics over the past two centuries, one is at once struck by the lack of interplay between the two disciplines. The number of theoretical economists who have had any sort of command of advanced mathematics has been extremely small, and, up until the last two decades, one can hardly name a top-rank mathematician who interested himself in problems of theoretical economics. The state of affairs is the more surprising when contrasted with the very close relation that has always existed between mathematics and the physical sciences.

It is not difficult to account for this discrepancy. Most economists agree that if a true science of economics is to develop, appropriate mathematical machinery must first be found. However, the economic problems of the moment are always of such urgency that there is tremendous pressure on the economist to get immediate results from data at hand using his intuition, experience, and whatever mathematics he happens to know. Mathematicians, on the other hand, have always been eager to attack problems in any field once they have been formulated in precise terms, but they are inclined to shy away from questions whose mathematical formulation is vague or incomplete, as has generally been the case in economics.

It is a cheering fact that during the last twenty years there has been some progress in bridging the gulf between economics and mathematics. We should rather say that a common ground between the two disciplines has been found on which the adherents of each can work without feeling entirely out of their element. We are referring to the study of what are called economic models. These models might be termed first approximations to an actual economic situation in which deliberate over-simplifications are made in order to obtain unambiguous and mathematically precise formulations of a problem. To be sure, this program of attack calls for abandoning, at least temporarily, the quest for immediate practical results or even experimental verification. There is, nevertheless, excellent historical precedent for such a program. Newton's first law of motion, that a body in the absence of external forces continues to move with constant velocity, cannot be verified by any terrestrial experiment, yet without the idealization of a frictionless universe the science of mechanics could not have been born.

At the present time there appears to be an ever-growing interest among both economists and mathematicians in the analysis of economic models. We mention a few of the events which have set the current trend in motion. In the middle thirties two major contributions by eminent

33

mathematicians appeared in the field of theoretical economics. The first was a series of papers by Abraham Wald in which the author analyzed the classical equations for determining prices and production levels at economic equilibrium. The equation system had been formulated at the beginning of the century but no one had previously investigated the nature of its solutions or indeed whether solutions existed. Wald was able to show, using mathematical methods of considerable sophistication, that under appropriate conditions solutions did exist and, equally significantly, he gave a simple example of a case in which the equilibrium equations had no solutions. At about the same time the mathematician von Neumann showed by means of a simple model how an economy could expand at a constant rate and how prices could be determined by this uniform expansion. Finally, we mention the well-known book, "Theory of Games and Economic Behaviour," by von Neumann and Morgenstern in which the authors propound the thesis that a competitive economy resembles a very complicated game of strategy involving many players. By carrying through a thorough analysis of these game models it was hoped that light might be thrown on the more complex economic situations.

We have listed only a few of the events that have led up to the present increasingly vigorous activity in the analysis of economic models, activity which has already produced some interesting results. Of course, the big problems, the "burning questions," are still unanswered and will surely remain so for some time to come. Progress has, however, been made in another direction, namely, in the development of mathematical techniques. In the examples cited above and in most of the present investigations it has been found that the type of mathematics needed is not at all like that used by the classical economists. Again, if history is any guide, this is an encouraging sign. The science of physics had to wait for the invention of the calculus, and one would reasonably suppose that a true science of economics waits upon the development of a mathematics appropriate to it.

We are about to turn our attention to some specific economic models. The models are simple enough so that the reader should find no difficulty in understanding them. We shall then try to give a very rough indication of the mathematical methods required for their solution. In this way we hope to convey something of the flavor both economic and mathematical of this current trend in theoretical economics.

A General Problem

In any economic system there are two important types of data which are determined outside the system itself and which the economist must accept as given. They are:

(1) The production possibilities.

(2) The tastes of consumers.

By the production possibilities we mean detailed information on what goods can be produced, in what quantities and by what processes: how much steel, rubber, labor, etc., is required to build one automobile; or how much of calories, vitamins, rest, is necessary to produce one laborer. This information can often be given quantitatively in the form of production functions or relations which tell in a concise manner what processes of production are technologically possible. Clearly, a knowledge of the production possibilities is essential for any realistic economic analysis. The question of how the production functions are determined, however, is not the concern of the economist but rather of the engineer.

Information on the tastes of consumers is much harder to express quantitatively, but just as important for economic analysis. Given a choice between two different goods, which will a consumer select, what proportion of consumers will prefer the first to the second? It is clearly essential, and usually very difficult, for economists to learn as much as possible about such questions. However, the theory of how these preferences are formed is again outside the field of economics itself, belonging, perhaps, to some branch of applied psychology.

Let us now assume that there is an omniscient economist who has complete information on both (1) and (2) above. He is then faced with two fundamental questions which are purely economic. Namely,

(A) What goods will be produced and in what amounts?

(B) How will these goods be distributed among consumers?

Note that problems (A) and (B) are universal in the sense that they are common to any economic system, be it one of free competition or complete central planning. Given the information (1) and (2), the economic system must determine the answers to questions (A) and (B). In quasi-mathematical terms, the independent variables are given in (1) and (2) and one seeks to find values of the dependent variables which answer the questions (A) and (B).

The models we are about to consider are idealizations of certain aspects of the general problem just described, and in studying them it is well to keep in mind their relation to the general case.

A Linear Model of Exchange

We consider first a simple model of exchange developed some years ago by the Norwegian economist Ragnar Frisch. The model can be best introduced by means of an example.

Let us imagine a very simple economy consisting of only three persons, a farmer, a tailor, and a carpenter. The only goods in the economy are therefore food, clothes, and housing. We further suppose that we know how each member of the group will distribute his income in buying goods from the other members. Assume, for the sake of concreteness, that out of every dollar he earns (from the sale of food) the farmer spends 50¢ on clothes and 50¢ on housing, while the tailor spends 25¢ on housing and 75¢ on food, and the carpenter spends 40¢ for clothing, 60¢ for food.

We now ask the question: given the information above, how much of each of the three goods will be consumed per day? Let x_1, x_2, x_3 represent the amounts in dollars' worth of food, clothes, and housing consumed daily (we ignore the amount which each tradesman consumes of his own good, since such consumption is assumed to be free and therefore does not affect the amounts of goods exchanged). The problem is solved by simply observing that the amount of each good sold by one of the tradesmen must equal the amount bought by the others. If x_1 is the amount of food sold by the farmer, and the total income to the tailor is x_2, he will buy $0.75x_2$ dollars worth of food, and similarly the carpenter will buy $0.6x_3$ dollars worth of food, and we obtain

$$x_1 = 0.75x_2 + 0.6x_3.$$

In the same manner, the amounts of clothes and housing sold must equal the amounts bought. In this way we obtain the following equations for the unknown amounts x_1, x_2, and x_3:

$$x_1 = 0.75x_2 + 0.6x_3,$$
$$x_2 = 0.5x_1 + 0.4x_3,$$
$$x_3 = 0.5x_1 + 0.25x_2. \tag{1}$$

The above situation is conveniently illustrated by means of the following simple table:

TABLE I

	Food	Clothes	Houses
Farmer	x_1	$-0.5x_1$	$-0.5x_2$
Tailor	$-0.75x_2$	x_2	$-0.25x_2$
Carpenter	$-0.6x_3$	$-0.4x_3$	x_3

In this table the quantity in the ith row and jth column represents the amount of the jth good sold by the ith tradesman, where selling a negative amount of a good means buying that good. Note that no matter what numbers x_i are places in the table, the sum of the terms in each row is zero. This is simply the requirement that each tradesman must spend his entire income in purchasing the goods of the others. Also the condition that the numbers x_i satisfy equations (1) is the requirement that the sum of each column in Table I be zero.

At this point we must issue a sharp warning. Observe that (1) is a system having the same number of equations as unknowns. On the strength of this fact can we conclude that our economic problem has a solution? The answer is a decided "no." To see how misleading such reasoning can be, note that the equations have the obvious solution, $x_1 = x_2 = x_3 = 0$, but this is hardly a proper solution to our economic problem since it implies that all consumption is zero. If this were the only solution of (1) it would mean that the members of the economy were condemned to slow extinction by cold, exposure, and starvation. We stress this point, for economists have often been guilty of drawing conclusions by the method of counting equations and unknowns. Here we see how fallacious such reasoning can be even in the simplest cases.

Clearly for a solution of (1) to be economically meaningful we must require that none of the numbers x_i be negative, and if the solution is to give a satisfactory answer to our problem we must also demand that not all of the x_i equal zero. Thus, we seek a solution of (1) satisfying the additional conditions

$$x_1 \geq 0, x_2 \geq 0, x_3 \geq 0,$$

and,

$$x_1 + x_2 + x_3 > 0. \tag{2}$$

There are now in all seven conditions on the three unknowns and the existence of a solution is no longer obvious. We are dealing here with a problem involving both linear equations and inequalities. The mathematical theory of such systems has been quite thoroughly developed but is neither so well known nor elementary as the ordinary theory of linear equations. Further, it has not in the past been part of the normal equipment of the theoretical economist. A great proportion of the current work in economic models involves exactly this branch of mathematical analysis, and it is becoming increasingly clear that a thorough grounding in this subject will be at least as important to the economist of the future as the calculus he already knows so well.

Returning to the concrete problem at hand, we can, by simple elimination, compute all solutions to equations (1) and we find that x_1, x_2 and x_3 are solutions if and only if they are in the ratios $36:28:25$. In other words, our problem does have a solution in which all the numbers x_i are positive.

Chapter Five

RELATION TO SOCIETY

Much public attention has been focussed on science and technology since World War II. The most important reasons, probably, stem from the contributions of science to the conduct of that war, especially the development of the atomic bomb and radar. Other factors are the population explosion, environmental pollution, and the cost of maintaining a science establishment with a balance of education, research, development, and production. Science as an institution is now truly an organic part of our society, not an isolated mechanism that can be turned on and off at will.

Along with this attention being given science, there is among some people a mystical faith that science can achieve anything desirable, while among others there is alienation and despair that science may lead to the destruction of mankind. Which of these expectations is correct? Both probably contain a germ of truth, but both also suffer from exaggeration rooted in lack of understanding. Robert S. Morison examines the attitudes of laymen toward science, attitudes that reflect the concerns mentioned above. He proposes some concrete steps, of which the publication of this book is one, for creating a better public understanding of the values of science. In the second article, I. I. Rabi addresses his fellow scientists and urges them to be more sensitive to their own role and the obligations it imposes. Mutual understanding, after all, is a two-way proposition.

What is your attitude toward science in general and physics in particular? Are you one of the optimists or one of the pessimists? And how do you suppose the public learns about the work of scientists? Formal education is one way, but the press is probably the most important medium through which scientists can communicate with the public at large about current events. The science writer and the editor serve key roles in selecting and interpreting newsworthy material, as is pointed out by Leon E. Trachtman and Allan R. Starry in their study. You may wonder about the results of this reporting effort when you consider that "newsworthy" items are those that will appeal to a public often in the grip of serious misconceptions.

In the emerging nations, science is viewed quite differently than in the United States, and it plays a different, more obviously constructive role. Abdus Salam presents an overview of progress and problems, with special emphasis on Pakistan. At the same time, he gives the matter an interesting historical perspective which will remind you that the West did not always have scientific leadership.

The last two selections refer to specific instances in which the work of physicists had direct public significance. One of them deals with the tragic assassination of President John F. Kennedy. Confusion and controversy have surrounded efforts to reconstruct the chain of events, because the happenings were observed poorly and eye-witness accounts conflicted. While no one may ever know what "really" occurred, Judith Goldhaber's account makes clear how physicist Alvarez used his technical knowledge to squeeze information from unlikely sources.

The other selection is a firsthand account in which Merle A. Tuve relates the origin of radar. This event falls into the classic pattern of how an irritating nuisance in Tuve's research later turned out to have inestimable value in war and peace. Radar, after all, has been an essential element in the expansion of air travel that has revolutionized transportation during the past twenty years.

In the future, you will probably get most of your information about science from the mass media. Take a look at your newspapers, listen to your radio station--what is the picture of science they communicate? Is science relevant to you?

About the Contributors

ROBERT S. MORISON (1906-   ) is Director of the Division of Biological Sciences at Cornell University. Professor Morison is a research physician and neurophysiologist who spent many years as administrator in the health sciences for the Rockefeller Foundation, working with Warren Weaver. He is now interested in the relation of science and technology to society rather than in studying biology itself.

I. I. RABI (1898-   ) became University Professor at Columbia University after many years of extraordinarily distinguished teaching, research, and government service. Professor Rabi received the Nobel Prize in physics in 1944 for his work on the magnetism of atomic nuclei.

LEON E. TRACHTMAN (1925-   ) is Associate Professor of English and Assistant Dean at Purdue University. Professor Trachtman is an active science writer and a teacher in this field. His experience has convinced him that a science writer must be a trained critic who is willing to evaluate the implications of science for society, not merely a "translator" who makes technical information available to the public.

ALLAN R. STARRY (1934-   ) is Associate Professor of Psychology and Director of the Measurement and Research Center at Purdue University. Although primarily an applied statistician, Professor Starry has personal and professional interests in the field of communications.

ABDUS SALAM (1926-   ) is Professor of Theoretical Physics at Imperial College of Science and Technology. To promote the advanced training and professional motivation of scientists in the emerging nations, he founded the International Centre for Theoretical Physics in Trieste, Italy, in 1964. Dr. Salam is also Chief Scientific Adviser to the President of Pakistan.

JUDITH GOLDHABER is a science writer and editor at the University of California's Lawrence Radiation Laboratory, where she edits The Magnet, a monthly nontechnical review of Laboratory activities. Her science background consists of a two-year interdisciplinary course required of under-graduates at Brooklyn College.

MERLE A. TUVE (1901-   ) is a Distinguished Service Member of the Carnegie Institution after serving as director of its department of Terrestrial Magnetism for twenty years. Dr. Tuve has won many awards and prizes in recognition of his work in geophysics and radio propagation.

Science and Social Attitudes

Growing doubts require that science be put
more recognizably at the service of man.

Robert S. Morison

Like all people with some scientific training, I suffer from feelings of unease when attempting to deal with the actions, and especially the attitudes, of people. For one thing, I do not have at my command the sampling and interview techniques wielded with so much aplomb by my colleagues in the social sciences. Fortunately for my own piece of mind, my scientific training was accompanied by enough exposure to the art of medicine so that I retain considerable respect for clinical intuition and judgment. This discussion relies much more on these elusive instruments than it does on quantitative scientific analysis.

As a matter of fact, it puts no great strain on one's clinical intuition to observe that large numbers of people in various parts of the world—including, perhaps most significantly, the advanced parts—are less happy about science and technology than they once were. The evidence is of various kinds. Perhaps the most quantitative is provided in the United States by the relative decline in students entering the sciences and the scientifically based professions. In some instances, such as engineering, the numbers have fallen absolutely in the face of a steady increase in the total number of

potential students in each age class. Even more quantitative, and certainly more compelling to the individual scientist, is the evidence provided by the slowdown in appropriations for science. Third, one may cite the intuitions and reflections of thoughtful social clinicians like René Dubos (1), who has so courageously summarized the shortcomings of scientific approaches to human problems. True enough, he finally draws the conclusion that what we need is not less science but more. Nevertheless, the argument depends on a careful demonstration that science raises new problems of increasing complexity as it continues to solve the older and simpler ones.

Earlier Attitudes toward Science

Before going on to a discussion of the possible reasons for a decline in public regard, we should pause to remind ourselves that the change may not be so large or so profound as we might suppose. It is not very clear that there ever was a time when a substantial part of the population really understood science, cared much about the kind of knowledge it produces, or thought much about its ultimate effects. Improvements in technology were welcomed because of the increased production of what were generally regarded as good things at less cost in human effort. On the other hand, the reduction in human labor was soon recognized to have a negative side. In the first place, as the Luddites saw very early, it tended to throw men out of work, at least temporarily. What was even worse from the psychological point of view, the machine tended to change the status of skills which had been acquired with much effort over long periods of years. Nevertheless, on balance, the industrialization of production both on the farm and in the factory has been regarded by most people as a net good; for, it must be remembered, even at the height of the Medieval and Renaissance periods, skilled craftsmen constituted only a very small portion of society. The great bulk of mankind labored in the most unimaginative and unrewarding way as farmhands with a status little better than that of serfs. Somewhat later than the general recognition of technological improvements in production came an even greater appreciation for the contributions of science to medicine and public health.

Most men probably never did take much interest in what might be called the philosophical aspects of science. Few really read Condorcet or the other Encyclopedists, and it is doubtful that any but a small handful of intellectuals ever thought that science would provide a way of life free of undue aggressions, anxiety, loneliness, and guilt. Perhaps the Communist Party is the only large social organization that has ever seriously believed that man himself may be improved through improving his material circumstances. Among Christians, as among adherents of many other religious faiths, there has always been a substantial body of opinion which holds that the reverse is true and that material prosperity has, in fact, an adverse effect on the human soul.

The progress of science undoubtedly has had some effect in reducing the grosser forms of superstition. One supposes, for example, that most men are in some sense grateful for being less afraid of thunder and lightning than man used to be. But, here again, it is doubtful that the scientific way of looking at the world has ever completely displaced older, more magical approaches to the deep questions. It does not appear that President Nixon, when making up his mind whether or not to deploy the ABM system, consulted an astrologer, but it is not unknown for heads of states in other parts of the world to do so, and most of our metropolitan daily newspapers maintain an astrology column as well as the more sophisticated services of Ann Landers. Indeed, it is estimated that there are 10,000 professional astrologers but only 2000 astronomers in the United States (2).

Putting aside the grosser forms of superstition and turning to better-developed and better-thought-out ways of looking at the world, I would hazard a guess that the metaphysical outlook of most people, even in the United States, is more influenced by Plato and Aristotle than by Galileo and Hume. Indeed, it might be interesting for a graduate student in intellectual history to survey this very question. For example, do you suppose the majority of Americans would consider the following statements to be true or false? "Other things being equal, heavy bodies fall faster than light ones." "Metals feel cold to the touch because that is their nature." "Justice and honesty are real things and part of the divine plan of the universe; men try to establish justice through the machinery of the law and the courts, but their efforts will always fall short of the higher ideal of justice as it exists in the divine plan."

Coming down out of the clouds, we might ask ourselves how many people ever really got much fun out of studying mathematics and physics in high school? How many felt pleased to discover that a suction pump doesn't really suck water, but merely creates a potential space into which the water is pushed by atmospheric pressure? If one looks back 40 or 50 years, one seems to remember that rather less than the majority of one's classmates really enjoyed physics and chemistry and the kind of picture they give of the world. Perhaps a somewhat larger number found satisfaction in biology, with its greater emphasis on immediate experience and the pleasure one gets from contemplating nature's wide variety rather than its unifying mechanics.

World War II called a great deal of attention to science and made many people grateful for its role in enabling England and its allies to maintain the integrity of the free world. Along with the extraordinary buildup of military technology came a very great increase in biological knowledge of a kind which could be applied to medicine and public health, and to agriculture.

The press showed increasing interest in reporting scientific events, and the quality of scientific reporting has greatly improved in the quarter century since the war. Most significantly, a grateful and more understanding public provided vastly increased financial support for what the scientist wanted to do.

On the scientist's side there was a burgeoning of interest in making science more accessible to the general public. Most noteworthy in this movement, at least in the United States, was the effort of outstanding university scientists to improve the presentation of science to students in elementary and secondary schools. There is little doubt that this effort has greatly improved preparation for college in all branches of science. The generous men who initiated the program hoped for something more, for they felt that, if the story could only be presented properly, anyone of average intelligence would share the pleasure of the most able scientist in discovering the orderly arrangement of the natural

world. Nothing could be more admirable than the dedication and self-sacrifice of men like Zacharias and the late Francis Friedman, and nothing more charming in its humility than their apparent belief that almost everyone is potentially just as bright as they themselves. Unhappily, it has not turned out as they hoped. Elegant though the Physical Science Study Committee Physics Course undoubtedly is, it has not proved much more successful than any other method in making physics attractive to secondary school students.

Nevertheless, on balance, public interest in science became greater after the war than it had been before, and it was further stimulated by the orbiting of Sputnik. It is very difficult to say how much of this interest was due to competition for ever more sophisticated weapons, how much to a pure cultural rivalry which puts the moon race into the same category as an Olympic track meet, and how much to the age-old wish to cast off the shackles which bind us to a single planet. However one apportions the credit among these three factors, it seems reasonably clear that an appreciation for basic science, as the scientists understand science, played a relatively small role.

Reasons for the Change of Mood

The decade of the 1960's has certainly seen a slackening in public approval of science. Is this change simply a return to the earlier, more or less normal state of ignorance and indifference, or are we witnessing an actively hostile movement? In either case it may do us all good to try to identify some of the more important reasons for the change of mood.

1) Science is identified in the public mind largely with the manipulation of the material world. It is becoming clearer and clearer that the mere capacity to manipulate the world does not insure

that it will be manipulated for the net benefit of mankind. Nowhere is this more obvious, perhaps, than in the matter of national defense. As pointed out above, the generation that knew at least one of the great world wars is grateful to the scientist for having fashioned the means of victory over a grave threat to a free world. The oncoming generation views the situation in quite a different way. To them the obvious alliance between the scientific community and the military is an evil thing: far from making the world more secure, it has produced an uneasy balance of terror, with the weight so great on both sides that any slight shift may lead to unimaginable catastrophe.

It seems undeniable that those of us who have grown up with this situation have also grown somewhat callous to the fact that such a high percentage of support for university science comes from military sources. We tend to remember, for example, the marvelously enlightened policy of the Office of Naval Research, which did so much to foster pure science while the Congress continued to debate the desirability of a National Science Foundation. Those who come upon the situation for the first time, however, see almost nothing but a conspiracy between some of the best brains of the country and the unenlightened military. In any case, it must be admitted that science and technology appear to contribute disproportionately to the more fiendish aspects of an evil business—the defoliation of rice fields, the burning of children with napalm, and the invention of new and more devastating plagues.

2) Until fairly recently, the contributions of science and technology to increased production both in industry and in agriculture have been generally regarded as on the plus side. Even here, however, doubts are beginning to arise. Much of the increased production comes at the cost of a rapid exhaustion of natural resources and the increasing contamination of what is left of our natural environment. Nor is it clear that all of the goods and services produced really do a great deal to increase the sum total of human happiness. Indeed, it can be shown that the modern affluent consumer is, in a sense, a victim of synthetic desires which are created rather than satisfied by increased production (3). On the other hand, a substantial percentage of the population remains without even the bare essentials of life. Rapid increases in agricultural production have pretty well abolished famine in the advanced countries of the world, but the revolution in rural life has benefited only a few of the most successful farmers. The rest are clearly worse off than they were before; and, indeed, the large majority of them are hastening into the cities, where they create problems which have so far proved insoluble. Furthermore, the advanced technologies which make the increased production possible are now found to be doing as much harm to the environment as the more long-standing and better recognized industrial pollutions.

3) Surely everyone can agree that science has done wonderful things for the improvement of health. But, even here, uncomfortable questions are being asked. Have our best doctors become so preoccupied with the wonders of their technology that they have become indifferent to the plight of large numbers of people who suffer from conditions just as fatal but much less interesting? Even the most earnest advocates of increased research in heart disease, cancer, and stroke must be a little bit embarrassed by the fact that the United States, which used to be a world leader in reducing infant mortality rates, has now fallen to 15th place.

4) It is not only the maldistribution of resources that concerns the general public; they are becoming increasingly uneasy about the moral and ethical implications of advances in biological science. In many respects these advances seem to threaten the individual's com-

mand over his own life.

Actually, of course, the individual never did have as much control over his own life as he felt he had. Science may have simply made his own impotence clear to him by showing how human behavior is molded by genetic and environmental influences. Like everything else, it seems, human behavior is determined quite precisely by a long train of preceding events, and the concept of free will has become more difficult to defend than ever.

Perhaps more immediately threatening is the fact that science puts power to control one's behavior in the hands of other people. Intelligence and personality tests place a label on one's capacity which is used from then on by those who make decisions affecting one's educational and employment opportunities. New methods of conditioning and teaching threaten to shape one's behavior in ways which *someone else* decides are good. Drugs of many kinds are available for changing one's mood or outlook on life, for reducing or increasing aggressive behavior, and so on. So far, these drugs are usually given with the cooperation of the individual himself, except in cases where severely deviant behavior is involved, but the potential for mass control is there. Indeed, there is already serious discussion about the ineffectiveness of family planning as a means of controlling the world's population, and suggestions are made for introduction, into food or water supplies, of drugs that will reduce fertility on a mass basis.

As if these assaults on individuality were not enough, some biologists are proposing to reproduce standard human beings, not by the usual complicated and uncertain methods involving genetic recombination, but by vegetative cloning from stocks of somatic cells. In the face of all this, can we blame the great majority of ordinary men for feeling that science is not greatly interested in human individuality and freedom?

5) Science is not as much fun as it used to be, even for its most devoted practitioners. The point here is that science encounters more and more difficulty in providing a satisfyingly coherent and unified picture of the world. The flow of pure scientific data is now so prodigious that no one can keep up with more than a small fraction of it. Although most of us still retain some sort of faith that the universe, with all its infinite variety of detail, can in some way be reduced to a relatively simple set of differential equations, most of us recognize that this goal is, in practice, receding from us with something like the speed of light. That simple set of physical and chemical principles on which the older generation grew up is now turning out to be not very simple at all, and the relation between these simple principles and the complex events of biology are not nearly so clear as they were when Starling enunciated his "law of the heart."

Although it is probably too easy to exaggerate the degree to which the progress of science results in the fragmentation of knowledge, the beginning student in the sciences finds a great deal of difficulty in relating his courses in chemistry, physics, and biology to one another. Even within a single discipline, he feels overwhelmed and frustrated by the number of apparently isolated facts that he has to learn.

6) Closely related to the foregoing thoughts on the growing complexity of science and the decline in the intellectual satisfaction generally derived from it is the question of student attitudes, for most of us make our first serious acquaintance with science as students.

My overall impression, in returning to a university after a lapse of 20 years, is one of disappointment that so few students seem to have very much fun either in their science courses specifically or in university life in general. This lack of pleasure is certainly more striking in the first 2 years, when the student is adjusting to a totally new social environment and devoting his attention to building the groundwork for later, more exciting studies. But I keep asking myself why these first 2 years of foundation-laying have to be so unsatisfying.

In the first place, I have come to believe that we discourage many students by expecting too much of them. We want them all to learn at a rate determined by the best. This can only mean that all *but* the best feel themselves to be dying of a surfeit rather than enjoying a marvelous meal. I am also coming sadly to the conclusion that, no matter how the subject is presented, a substantial number of college-level students have relatively little interest in the facts of science and lack the capacity to find pleasure in its generalizations. Whether the failure is primarily intellectual, in the sense that students simply have difficulty in understanding the nature of the generalization, or whether it is emotional and esthetic, in that they derive little pleasure from the generalization once it is understood, is not easy to determine. In either case, the prospect of unifying the community around a common understanding of science seems relatively remote.

An article by Richard N. Goodwin in the *New Yorker*, entitled "Reflections—sources of the public unhappiness" (*4*) puts some of the difficulties of science into a larger perspective. It provides a brilliant analysis of the unhappiness not only of our obviously dissident left-wing youth but of the many members of the forgotten middle class who, during the last election, swung rather wildly between George Wallace and Eugene McCarthy. Goodwin discusses this phenomenon in terms of the traditional Jefferson-Hamilton model and comes to the conclusion that a great many Americans feel that they have lost control of certain crucial factors in their life styles. Although I am far from being as convinced as Goodwin is that it will be possible to return a large portion of our decision-making to states and local communities, I agree with much of his analysis of the underlying problem. He is particularly convincing, for example, when he shows how Secretary McNamara, in his apparent efforts to rationalize the Department of Defense and bring the military more closely under civilian control, actually succeeded in constructing a Frankenstein monster which began to control him, as "when he was compelled against his own judgment to go ahead with an anti-ballistic missile system."

For our purposes, the key word here is "rationalize." Our rationalized systems do, indeed, seem to have developed the capacity to live lives of their own, so that mere men are compelled, against their will, to follow where the logical process leads. As we saw above, the medical profession is following in the footsteps of its dynamic research program and undertakes to perform heart transplants, at great expense, largely because it has found out how to do them. In the same way, we devote several billions of dollars each year in going to the moon, because it is *there* (and, again, because we know how to do it). Everyone who has done much science on his own knows that the next step he takes is determined in large part by the steps that have gone before. It follows that the progress of pure science, at least, is determined by the internal dynamics of the process and by the opening up of new leads rather than by public demand to meet new needs. The practical applications to human welfare, when looked at in this philosophical framework, become accidental bits of fallout, as the nuclear bomb itself "fell out" from the innocent effort of J.,J. Thompson, Rutherford, Bohr, Fermi, and others to understand the nature of matter. No doubt all these men felt completely in command of their own research programs, but the public does not look at it this way, and, in a curious sense, the public may be more right than the scientists. This line of thought brings us to point 7 in our bill of indictment.

7) The continuing momentum of science toward goals of its own choosing appears to be coupled ever less closely to solving problems of clear and pressing consequence to human welfare. As we now see, enlightened congressmen and senators, well aware of the power of the scientific method but skeptical of its capacity to guide itself automatically to the points of greatest human concern, are making explicit legislative attempts to mobilize science to solve the problems of the pollution of our environment and the crime in our cities, if not, indeed, the unsatisfactory nature of our life in general. Realizing that nuclear physics is not very closely coupled to these matters, they are turning to social science in the hope that there is a group of scientists who can do for society what the physicists have done for the natural world.

Skepticism about Rational Systems

Skepticism about rational systems is, of course, not confined to science. Indeed, it well may be that the antipathy to science is merely a bit of fallout from the growing antipathy to rational systems in general (5). The movement has been a long time in the making. Lionel Trilling (6), for example, traces much of the despair, the irrationality, and the increasing devotion to the absurd of much modern literature to Dostoevski's *Letters from the Underworld*, in which, you will remember, the protagonist, in his violent diatribes against the existing order, concentrates his hatred on those "gentlemen" who believe that 2 and 2 make 4. What is even more frightening for our own time is the way the same anti-hero reassures himself of his own individual freedom by affirming his ability to choose the more evil of two options (7).

We, who have grown up rejoicing in science, were confident in our acceptance of Sir Francis Bacon's aphorism that we cannot command nature except by obeying her (8). We really did not mind obeying as long as we knew that we would ultimately command. But now the empirical evidence may be turning to support those who feel that science is in some sense in the grip of natural forces which it does not command. Too often we conjure up genies who produce short-term benefits at the risk of much larger long-term losses. We develop marvelous individual transport systems which poison the air we breathe; learn how to make paper very cheaply at the cost of ruining our rivers; and fabricate weapons that determine our defense strategy and foreign policy rather than being determined by them. Above all, the applications of science have produced an unrestricted increase in the human population which we recognize as fatal to our welfare but have only the vaguest idea how to control. In a short time we will be able to design the genetic structure of a good man. There is some uncertainty about the exact date, but no doubt that it will come before we have defined what a good man is.

In the foregoing analysis, in an effort to obtain intellectual respectability I have painstakingly tried to break our problem down into a series of numbered subheadings. Actually, they all add up to the same thing: Although the general public is grateful to science for some of its more tangible benefits, it is increasingly skeptical and even frightened about its long-term results. The anxiety centers on the concept of science as the prototype—the most magnificent and most frightening example of the rational systems which men make to control their environment and which finally end by controlling *them*. It may be well to recall that the medieval structure of natural law was even more rational than science, in the sense that it depended on the mind alone without submitting its conclusions to empirical checks. It managed for a time to obtain even greater control than science has over both the bodies and (especially) the spirits of the people of the Western world. It, too, developed an interesting life of its own as it followed the paths of reason into ever more subtle areas. It failed, for a number of reasons, but primarily, perhaps, because neither the logic-chopping of the medieval philosophers nor the temporal power of the papacy which it was designed to support appeared to be sufficiently related to the longings of individual human beings. The Reformation, for all the complexity of its theology and, often, the brutality of its methods, was primarily an effort to assert the rights of the individual conscience over the medieval power structure.

A Watershed?

I am not really sure that we stand on the kind of watershed Luther stood on when he nailed his theses to the door of the cathedral, but we may make a serious mistake if we do not at least entertain that possibility. If we fail to recognize the average man's need to believe that he has some reasonable command over his own life, he is simply going to give up supporting those systematic elements in society which he sees as depriving him of this ability.

As I noted above, so perceptive a critic as Lionel Trilling traces much of modern literature and art to a long-standing revolt of sensitive and creative men against the systematic constraints of society. The New Left can be regarded as a politization of the same trend. Actually, of course, anarchy had a political as well as a purely intellectual existence when Dostoevski was writing, but the 19th-century political anarchists were effectively liquidated by the Marxists, who felt that they had a better idea. Now that Marxist communism has developed most of the ills of bourgeois industrial society plus its own especially repressive form of bureaucracy, anarchism is again put forward as an attractive

alternative to organized, corrupt societies.

There is a difference, however, in the way 19- and 20th-century anarchists regard science. On the whole, the 19th-century ones were atheists and saw religion as the co-conspirator with government and business. Science tended to be favored, partly because of its contributions to man's material welfare, but perhaps even more because of its aid in debunking religion.

Two paragraphs from Mikhail Bakunin are worth quoting, partly because of the flavor of the rhetoric (9).

[The churches] have never neglected to organize themselves into great corporations . . . the action of the good God . . . has ended at last always and everywhere in founding the prosperous materialism of the few over the fanatical and constantly famishing idealism of the masses.

The liberty of man consists only in this: that he obeys natural laws because he has himself recognized them as such, and not because they have been externally imposed upon him by any extrinsic will whatever, human or divine, collective or individual.

The New Left certainly agrees with Bakunin about the need to destroy the existing order, but it tends to see God in a different light. In the United States, religion has been conscientiously separated from the State for so long that it is no longer regarded as part of the apparatus of repression. Indeed, many draft-card burners and other protestors against the immorality of the existing order are primarily religiously motivated. On the other hand, science as the interpreter of the laws of nature, which Bakunin set against the laws of the State, has lost its revolutionary character and is viewed as a dangerous collaborator of the industrial-military complex. One of the difficulties may be that science has become so complicated that the ordinary man no longer believes that "he himself has recognized them [natural laws] as such" but feels that "they have been externally imposed upon him."

Educating the Public

What, then, can we do to improve the image of science as something of human scale, understandable and controllable by ordinary men? In the first place, we will have to continue our efforts toward educating the public, both in school and outside it, through reporting in our newspapers and magazines. Although I have given some reasons for believing that there are limitations to the capacity of much of our population to understand and take pleasure in the way science understands the natural world, I still believe that much more can be done to improve matters than has been done so far. As for the formal part of education, I propose that we rather deliberately reduce the rate at which students must handle the material set before them, so that they can master it without feeling frustrated and overwhelmed. If we begin the process, as is now fashionable, in the early elementary years, continue it through college, and carefully design things so as to avoid redundancy, students might end up with a much more complete understanding than they do now. This effort is worth even more money and time than have been put into it so far.

As for less formal methods for presenting science to adults, we should devise some analogy that would do for the general public what agricultural extension courses have done for the farmer and his wife. The average successful farmer, although he is far from being a pure scientist, has an appreciation for the way science works. Certainly he understands it well enough to use it in his own business and to support agricultural colleges and the great state universities that grew out of them.

As one who has spent a considerable period of his life worrying about medicine and public health, I am much less happy about our efforts to instruct the average man in a rational or scientific attitude toward the conduct of his own life. It has proved ever so much easier to persuade the average farmer to plant hybrid corn than to persuade the average man to give up smoking cigarettes. We have been almost too successful in persuading farmers to put nitrogen on their fields, while we continue to fail in trying to persuade the average man to put minute amounts of fluorine in his water supply. Few individual doctors seize the opportunity to explain to their patients, in even quasi-scientific terms, what their illnesses are, and I am appalled by the bizarre notions of human physiology which are entertained by some of my best friends.

Granted that doctors do not have enough time to talk to their patients and that many doctors really are not very scientifically oriented themselves, we might think seriously of setting up in every city a kind of paramedical service designed to teach people about their own illnesses. A doctor with a patient who is developing coronary insufficiency, for example, could refer his patient not only for an electrocardiogram, a blood-cholesterol, and clotting-time determinations but for instruction, in a class of cardiacs, on just how the heart and circulation work. Such an enterprise might help individual patients adjust to their illness more suitably, but this is not the real point. The aim would be to take advantage of an unhappy accident in order to increase the individual's motivation to learn something about science. Therefore, such clinics should be paid for not only by the Public Health Service but by the Office of Education.

Second, we must make a major effort to bring the course of science, and especially its technological results, under better and more obvious control by individual human beings and their representatives. We are, it is true, slowly gearing ourselves to do something about pollution of the environment, but the overall guidance and control of this effort is largely in the hands of part-time experts who fly in and out of Washington to attend meetings which issue prophesies of doom or unsupported reassurances, as the composition of the particular panel may dictate. Somehow, thinking about the long-term results of technology, formulating the options in such a way that the public can understand them, and guiding the course of events along the chosen path must become as exciting and rewarding for the best minds as is the present pursuit of basic scientific knowledge. Above all, the options must be made clearly understandable to the people, and the people must feel that they are doing the choosing. The present method of announcing that such-and-such a corporation is about to erect a large atomic power plant on a certain body of water and then engaging in a debate, based on inadequate information, about the effects of the heat on the lake or river, the degree of radioactive contamination, and so on, is totally unsatisfactory.

The process of educating the public should begin much earlier, with discussion of the need for additional power plants and of the probable cost of putting them here or there, in terms of increased power rates on the one hand and increased contamination of the environment on the other. The public must slowly be brought to see that every such occasion involves a real choice between real alternatives, and that the alternatives must be balanced against one another. Similar considerations apply to the use of insecticides. Nobody, as far as I know, has seen fit to make any even approximate estimates of what our food might cost if we were to abandon the use of these agents. Similarly, nobody has told what it would cost to produce high-octane gasoline by means of some method other than the addition of tetraethyl lead.

We have been very negligent in devising ways and means of ensuring that the cost of introducing new technologies is borne by the people who immediately benefit from their use. If anything, the trend may be away from emphasis on this relationship. For example, the introduction of the cotton picker and of modern methods of weed and insect control, not to mention the enormous subsidies provided by the American taxpayer, have made the culture of cotton in a few counties in the South and Southwest extremely profitable, so that large landholders have become extremely wealthy. Presumably, the public at large has benefited by a slight reduction in the cost of cotton cloth. On the other hand, the social costs of this industrial revolution in agriculture have been incalculable; they have been borne primarily by the large number of Negro laborers who have been uprooted and transported into the cities, where they found themselves ill-prepared to benefit from the urban amenities enjoyed by their more prosperous fellow citizens. The economic costs of supporting them in an alien environment have been borne, not by the wealthy southern landowners and certainly not by the individuals who paid a bit less for the cotton cloth, but almost entirely by the displaced people themselves and by the people who pay real estate taxes in a handful of our larger cities.

All these problems are, however, subject to some kind of scientific analysis, and the options can be placed scientifically before the public. In preparation for this kind of decision making, we should probably overhaul our teaching of science, and especially of mathematics, so as to give the average man greater ability to evaluate evidence presented in modern scientific form. High school courses in statistics, probability, and systems analysis are clearly more relevant to modern living than Euclidean geometry, and might well replace this and other time-honored introductory courses in mathematics.

Role of Science in Military Affairs

Third, an effort should be made to clarify the role of science in military affairs. Although most of us who are acquainted with the facts know that much of the research supported by funds from the military services actually contributes as much to civilian life as to military matters, this fact is not known to the general public or to the student body. Cornell students, for example, are disturbed to learn that the largest single donor to research at their university is the Department of Defense, even though one of the university's two largest research enterprises is the observatory at Arecibo, whose contributions to pure science are of far more consequence than anything it has ever contributed to the Air Force. If the military uses of science occurred as fallout from scientific investigations undertaken for peaceful purposes, this would be far better for morale than continuation of our present course, in which pure science appears as the crumbs that fall from the rich Pentagon's table. The obvious and actually very easy way to accomplish this would be to reduce military appropriations by what, to the military, would be a tiny amount and substantially increase appropriations for the National Science Foundation and the National Institutes of Health. Certain civilian agencies, such as the Department of Commerce and the Department of the Interior, should also be supporting far more basic and applied research than they are now.

Whether the civilian establishment for science should engage in any research of military consequence is a matter for debate, but such debate should be encouraged. Many universities of good will long ago decided that secret research has no place on a university campus, but this does not prevent them from doing unclassified work which has a clear military bearing, nor does the university ordinarily discourage its faculty from serving as consultants on classified projects carried out elsewhere.

There are obvious theoretical and practical difficulties confronting any other policy. Until now, for example, most scientists have felt that the importance of advancing knowledge overshadowed questions regarding the source of support. The control we now have of malaria is a net gain, regardless of the fact that, from the discovery of the malarial parasite in North Africa to the development of control methods by the American Army during World War II, research on malaria was often carried on by military personnel.

Furthermore, it is clearly important that we have, as consultants to the military, civilian scientists who learn the details of proposed weapons systems so that they can make an appropriate case against, as well as for, deployment of these systems.

Finally, as long as we feel ourselves threatened by the scientific and military establishments of other nations, it is with some difficulty that most of us who have special skills, gained largely through contributions from the American public, can refuse to use those skills for the defense of that same public. This last issue is becoming a rather knotty one, however, since we may have reached a point at which war is so disastrous for both sides that there is simply no point in undertaking the exercise at all.

Conclusion

The most important lesson for the scientific community would appear to be one that can be stated as follows. Science can no longer be content to present itself as an activity independent of the rest of society, governed by its own rules and directed by the inner dynamics of its own processes. Too many of these processes have effects which, though beneficial in many respects, often strike the average man as a threat to his individual autonomy. Too often science seems to be thrusting society as a whole in directions which it does not fully understand and which it has certainly not chosen.

The scientific community must redouble its efforts to present science—in the classroom, in the public press, and through education-extension activities of various kinds—as a fully understandable process, "justifiable to man," and controllable by him. Scientists should also take more responsibility for foreseeing and explaining the long-term effects of new applications of scientific knowledge. A promising procedure for planning the control of such effects is presentation of the probable outcomes of various available options so that choices can be made by the public and their representatives. Costs and benefits must be estimated not only in quantitative, dollar terms but, increasingly, in terms of qualitative and esthetic judgments. Thus ends the comfortable isolation of science from the ordinary concerns of men as a "value-free" activity.

References and Notes

1. R. J. Dubos, *The Dreams of Reason* (Columbia Univ. Press, New York, 1961), p. 167.
2. C. E. Sagen, personal communications.
3. J. K. Galbraith, *The Affluent Society* (Houghton Mifflin, Boston, 1958), chap. 3.
4. R. N. Goodwin, *New Yorker* 1969, 38 (4 Jan. 1969).
5. C. E. Schorske, "Professional ethos and public crisis: a historian's reflections," *Mod. Language Ass. Amer. Publ. 83*, 979 (1968).
6. L. Trilling, *Beyond Culture* (Viking, New York, 1965).
7. F. Dostoevsky, *Letters from the Underworld*, C. J. Hogarth, Trans. (Dutton, New York, 1913). All of part 1 is relevant to this discussion, especially page 37: "Moreover, even if man were the keyboard of a piano, and could be convinced that the laws of nature and of mathematics had made him so, he would still decline to change. On the contrary, he would once more, out of sheer ingratitude, attempt the perpetration of something which would enable him to insist upon himself; and if he could not effect this, he would then proceed to introduce chaos and disruption into everything, and to devise enormities of all kinds, for the sole purpose, as before, of asserting his personality. . . . But if you were to tell me that all this could be set down in tables—I mean the chaos, and the confusion, and the curses, and all the rest of it—so that the possibility of computing everything might remain, and reason continue to rule the roost—well, in that case, I believe, man would purposely become a lunatic, in order to become devoid of reason, and therefore able to insist upon himself."
8. F. Bacon, *Novum Organum* (1620), aphorism 129.
9. M. Bakunin (Bakounine), *God and the State* (1893).

A Matter of Opinion

O ne of the unhappy facts of our time is that while the new-generation scientist has made great gains in technical competence he has lost ground in philosophical understanding of his subject. We all carry in our minds a picture of what a true scientist should be—a person who wishes for nothing more than to understand nature in its broad sweep, its depth and originality. Although he himself may never be able to make a significant contribution, he is happy just to contemplate the grandeur that is science.

In contrast, the scientist of today is becoming more and more a specialist who doesn't take time to appreciate, or maybe doesn't even understand, the significance of his work in a global sense and its relation to man's aspirations. As a result, we find the scientists of today competing with one another to some degree on the level of technicians. Admittedly, as technicians they are on a very high level, often brilliant, but just the same their interests tend to be focussed along rather narrow technical lines.

The most immediate reason this problem exists is that the schools where our younger scientists have been trained no longer bother to foster appreciation of the universal and philosophical implications of science.

This is especially true of the graduate schools— graduate programs are very, very narrow. These days people think that if you add a course in the humanities to a graduate program you have really done something. But the real problem is that science itself—in the classroom lectures—is not being taught humanistically. When I say humanistically, I don't mean bringing in the Greeks, the Renaissance or things of that sort; I mean teaching science from the standpoint of its meaning to people of this particular time—its philosophical implications, its social implications, its cultural implications—all the things you need to know to understand why science has become the modern stage of human endeavor. You can't expect that science undergraduates will work this out on their own. A few might but most don't.

The disciplines themselves these days seem to encourage specialization. When I was a boy the philosophy department was queen of the campus and now philosophy itself has become so specialized into different groups that even philosophers hardly understand one another. The history-of-philosophy man doesn't talk very much to the man who is doing ethics or the man who is doing aesthetics or logic. This trend will continue until the universities become conscious of this problem and devote themselves to establishing groups or departments or schools of interdisciplinary study. The universities will have to do this because this sort of breadth is not encouraged by the disciplines themselves.

Suppose you want to have a professor of history of science. The science departments don't want him —they would rather use the money to get another scientist in their specialty. The historians couldn't care less about having him—they want another man in American history or medieval history. It's common to all disciplines that people don't appreciate a man stepping out of it for something else. Somehow or other they feel, often properly, that it is unsound. So there's very little room for interdisciplinary undertakings and so far universities are contributing to this specialization. Instead of working to broaden out the culture they are participating in chopping it up into bits.

Instead of more specialized courses we need more courses like the one I have been teaching for the last few years at Columbia. It was called the Cultural and Philosophical Implications of Twentieth Century Physics. It was not a course in physics, culture or philosophy but considered the conjunction of these things and tried to show the breadth and importance of science to our modern way of life.

That universities are beginning to become aware of the problem is evidenced by their recent increased interest in the history of science. The new interest in this subject should be very useful unless people make a profession of it and exclude everyone but the specialists.

The big loser with this specialization in science has been the general public. Most people live their lives without realizing they are living in the greatest age in the history of man. The reason they don't understand this is that they don't understand science and how it is changing our culture and producing the ideas and knowledge that provide direction for everything we do these days. They get only third- or fourth-hand knowledge about these things and instead of being interested and taking part in this adventure they feel more like a canoe thrown around in a torrent. As a result, most educated people have very little connection with the times in which they live—and it is no wonder that they feel alienated. Their education has not enabled them to understand what is really going on in the world.

You have a great clamor for more humanities in our schools but the humanity courses people talk about are rather meaningless in the sense that they don't talk about the present times, which are rather different from the ages of the classics. The young students ask, and rightly so, "It's okay about Socrates but how about Stokely Carmichael?" When we give this rather meaningless education to young people, they are unhappy—as anyone can see— especially the bright ones.

The trouble is our approach to education is in the tradition of the education of an English gentleman a century ago. We need to think about what an education really should be for young Americans to go out and live in *this* time. Our universities feel successful because when a boy comes in at 19 and goes out at 22 he has more often than not matured in many ways. But I wonder if the universities aren't taking credit for what association with other people and our magazines and newspapers have really done.

A young person in school now is going to be living maybe half his life in the twenty-first century. What are we teaching him to help him live there? Very little! What we need basically is a new look at the objective of education. The objective should be to develop people who can live in their age, understand the problems of their age and enjoy the great things that are happening.

And the most important subject we can get across to people is the role science has in shaping and unifying our 20th century culture. But we scientists ourselves are losing sight of the broad scope of things and the result is that instead of being the unifying element that could make some sense out of our fragmented culture, science has become just another fragment. As a group of specialists we have excluded the non-specialists—the general public—from our company. But the public is entitled to share the joys and excitement of science—after all the public is paying for it, for one thing, and, secondly, it is too good to keep to ourselves.

Unfortunately, we don't have scientists who write well about science in popular terms in the sense that Galileo did. Probably the reason is that scientists have not had to defend themselves as Galileo did. But now with science every year asking for a bigger slice of the national budget there has been a greater effort on the part of some disciplines to explain to the public what they are doing. Perhaps this economic pressure will be one of the things that motivates scientists to pay more attention to the social significance of their work.

Also there is one special group of scientists that hopefully we can count on to invest their energies in these problems. These are the scientists who have gone into business for themselves with small companies, such as you find on Route 128 outside Boston, and who in many cases have made a lot of money. I predict that we will find these financially independent scientists going into politics or taking influential goverment posts where they will be in good positions to spread "scientific culture."

Science in the press:
Black-and-white or gray all over?

by LEON E. TRACHTMAN *and* ALLAN R. STARRY, *Purdue University*

When the scientist and the newspaper science writer meet at a panel discussion, usually called "Science, the Mass Media and the Public," the scientists invariably accuse the writers of oversimplifying, distorting and misrepresenting scientific research. And the writers usually accuse the scientists of being elitist, exclusive and indifferent to the public's stake in their research.

As the discussion progresses, however, both groups discover they have a mutual scapegoat in the editor, the man who sets policies and decides which stories will appear in the paper.

Many editors invite this sort of attack by insisting that all science news tell, in a lead of 25 words or less, the who, what, where, when and why of the scientific development as well as its implications for the reader and society.

The occasional editor included on a panel defends himself, saying that his job is to sell papers and to do that he must print what the public will read, not what some scientist thinks it should read.

The implications of this debate are serious. Public attitudes toward scientific research, which are fostered by the mass media, will ultimately dictate the degree of support government will grant science.

Because of this, we decided to study science writing in a cross section of American newspapers. We wanted, first, to get gross figures on the volume and scope of science coverage in the daily press. Second, we wanted to determine whether this coverage accurately reflects the kinds and amounts of scientific activity actually being conducted.

We reviewed every issue of 34 daily newspapers for three months. We clipped, classified and analyzed every article relating to science.

Our most striking finding was that for all 34 papers an average of just under one science story was printed per day. (Papers with over

222

500,000 daily circulation printed between one and two stories a day, while those with under 40,000 daily circulation averaged only one science story every four days.) By any standards, this is grossly inadequate science coverage for a society as suffused with the attitudes and products of science and technology as ours is.

We first measured the reporting of developments in 10 disciplines in the scientific literature, because progress cannot truly be said to be made unless it is reported. This, to us, was the only objective way of measuring how much return society was getting from its investment in each discipline.

We then thought that there should be a discernible relationship between society's investment, the results of that investment (as measured in volume of science journal literature) and the newspaper coverage given to scientific achievements.

We used the disciplines that the National Science Foundation uses in its reporting of federal research expenditures: agriculture, biology, chemistry, earth science, engineering, mathematics, medicine, physics and astronomy, psychology and social science. We also used NSF's figures for total dollars obligated for these disciplines.

With help from the Institute for Scientific Information, a Philadelphia publisher of international indices to scientific literature, we developed figures showing the approximate number of scientific articles by discipline that appeared in American scientific journals. We then compared the expenditures and the number of journal articles with our figures on the newspaper reporting of scientific progress.

We found much disproportion. For example, engineering, where 33 percent of the dollars were spent, had only 12 percent of the scientific articles. Engineering also had a 14-percent share of newspaper coverage.

Medicine, on the other hand, got only 18 percent of the research dollars, but had 34 percent of journal articles and received 39 percent of the newspaper coverage. Chemistry suffered a graver disproportion in newspaper coverage. Here 8 percent of the expenditures produced 13 percent of the scientific articles but only 2 percent of the newspaper stories.

How do we explain these figures and what do they imply for science and scientists, for the makers of public policy and for the mass media?

First, only 15 percent of the members of the National Association of Science Writers actually work for the mass media. The rest work in public relations and public information offices, producing the news releases and brochures funneled to the mass media. This suggests there is just too much information, a good deal of it self-serving and scientifically insignificant, being sent to science writers and editors and that it is becoming nearly impossible for the media men to make a really balanced news selection.

Walter Sullivan of *The New York Times*, for example, reports that every day he receives a stack of releases between one and three feet high. Ninety-nine percent of these must be rejected rapidly. The problem of reading, screening and selecting for accuracy, timeliness and significance, in addition to meeting a daily deadline, inhibits the editor from balancing his coverage.

Editors must also decide what really constitutes science news. On one paper, the *Atlanta Journal*, city editor John Crown considers the advice-from-the-doctor column the paper's most important science coverage.

Most non-science news develops in the form of relatively discrete events about which spot stories can be written and on which in-depth, feature treatment can be hung. Not so with most science news. Scientific progress is normally made slowly and occurrences such as the first heart transplant are the exceptions.

The science writer, exposed to great doses of inconclusiveness, qualification and reservation, may weary of reporting research in progress and tend to select and emphasize stories that have a sense of completeness. This sort of story occurs with greater frequency in some disciplines than in others.

The scientific meeting is an additional source of many articles on science. But even here the writer is limited in choosing his subjects by screening committees or public relations representatives who preselect for distribution in the press room perhaps 10 percent of the papers being delivered at the meeting.

So, because of great volumes of material available and the special character of science news and its sources and the pressure of time, the newspaper tends to let outsiders select a high percentage of its subjects. Frequently the selection is honest, fair and unbiased. Occasionally, it is not.

Another factor in the imbalance of coverage is that much research today is being done under sponsorship of the Defense Department or private corporations and is restricted from publication. This is particularly true in engineering. Also, the inherent difficulty of explaining certain disciplines undoubtedly inhibits their coverage. Mathematics, for instance, received no coverage at all in our survey.

Another factor is the small number of newspaper science writers. Perhaps half of them are assigned full time to medicine and many are restricted by their own inadequate background or special interest in areas they can cover competently.

Finally, editorial judgment about what will appeal to the mass reader is a most important consideration in story selection.

To what degree is all of this bad?

The one real danger we see is that, because of undue concentration on certain disciplines and certain types of science news, the public may get a distorted view of science and of its role in modern life. A public educated to a "Mr. Wizard" concept of science will very likely develop false expectations and be frustrated when they are not realized.

The scientist, we believe, must decide to play a more active role in interpreting himself and his profession to the public through the professional science writer. He should support a growing trend in some newspapers to print less spot news on science in favor of more interpretive treatment.

Fearing misinterpretation and distortion, he should not, as so many scientists have in the past, shun the reporter, retreat to his laboratory and console himself by saying that his work is simply beyond the average man's comprehension. Much scientific work may be couched in language foreign to the layman, but the scientist should be able to explain the reasons for doing the work he does and the logical structure of that work and some of its implications for society.

Congress continues to question the utility of public support for certain areas of science. Certain newspapers editorialize against the government's supporting projects with apparently trivial or esoteric titles. Commentators criticize whole areas of scientific research as irrelevant to the critical problems of our society and suggest that such areas be less generously supported with public funds.

The scientist cannot continue to justify public support for his work on the faith that, as Murray L. Weidenbaum of Washington Univ. recently put it, "through serendipity . . . it will turn out to be worthwhile after all."

If scientists want public opinion to play a meaningful role in determining public policy toward science, they must actively participate in educating the public. We need a new kind of specialist—a scientist-journalist, who can give the public thoughtful analyses of scientific and science-policy matters.

Not all segments of society will ever be interested in science news. We should certainly not distort or oversimplify information in order to appeal to those segments, just as we do not oversimplify the complexities of the sports page for readers not interested in sports.

But the scientific community must reach the interested layman with thoughtful, critical and well-balanced science information, for it is this layman who is ultimately the prime shaper of our society.

How scientists feel about the press

—and vice versa

THE SCIENTISTS:

J. Alan Heineke, Northwestern University: *"There is a big problem with science coverage. Science is just not spot news and pure science appeals to a limited group of readers. Logically, a newspaper has to appeal to a large body of readers, so the selection of stories is usually based on what is of interest to the public. Sensational stuff sells.*

"They [science reporters] attempt to couch scientific feature material in spot news terminology. You just can't write a feature on a scientific program and make it read like a news flash about Vietnam. You lose the essence of the whole thing. They [the reporters] may not mean to make mistakes, but in many technical areas, no matter how knowledgeable they may be, there are always certain things they are just not aware of."

Heineke also complained that the press is not interested in the efforts behind research or the teamwork aspect of it. They want quick answers, he says, such as It is or It isn't.

Robert Dicke, Princeton University: *This physicist concedes that reader interest must dictate to a great degree what the newspaper prints but feels that the "better newspapers, as a means of education, should be providing a limited amount of news in certain scientific areas that are important to the formation of national science policy—whether the general public is interested or not."*

Arthur Kornberg, Stanford University: *This Nobel prize-winning biochemist feels that "lately the quality of reporting has been quite good," and that most science reporters are well qualified to cover the field. However, he complained about headline writers: "Cancer has a habit of working its way into the headlines no matter what role it plays in a story." He cited a recent biochemistry article that stated, deeply buried, that a discovery might influence the study of abnormal growths such as cancer. The headline read: "Major Breakthrough in Cancer Research." "Sometimes this kind* of headline prompts letters from relatives of cancer patients asking for help when there is really only the barest thread linking the two subjects."

Donald Harrison, Stanford University: "There is so much sensationalism. What is considered old hat to most acquainted with a field is played up into something to sell newspapers. Generally, science writers do a reasonably accurate job. However, nonscience reporters do a lousy job and are usually guilty of misemphasis and misquoting."

Barry Commoner, Washington University of St. Louis: "Generally the coverage has been best where the significance of research was self-evident. Subjects of controversy tend to be neglected. Some reporters tend to go with the majority opinion expressed by controversial situations. In my view this is not complete reporting. No matter who turns out to be right, controversy is an important aspect of the subject matter in the same sense as the standard deviation is an important part of any numerical datum." He said that the American press does not emphasize the uncertainties of science and that "the people need to know about the risks as well as the benefits. What we are lacking is journalistic critics of science who would operate as do literary critics. However, it would be even more important for critical discussions to occur within the scientific community. This is also lacking."

THE SCIENCE WRITERS:

Gobind Lal, San Francisco Examiner: "Scientists are just as good a news source as any other. It depends on the person. Of course, you must win their confidence if they are to really open up and tell you what they're doing. Some are jittery and want to shrink away."

Jerry Curry, St. Louis Post Dispatch: This reporter said he finds scientists cooperative 99 percent of the time, but likened them to careful politicians, extremely conscious of being misquoted. He said his science beat includes local hospitals, as well as the universities and research centers. Writing in depth is often hampered by lack of time, he said.

Art Snider, Chicago Daily News: This 23-year veteran of science journalism complains that many scientists "consider the popular press yellow journalism" and feel that dealing with newspapers is beneath their dignity. "Then there are the ones that never have the time to talk. They are always filled with excuses and are usually the ones who scream the loudest when they do want coverage and can't get anyone's attention." However, he feels that the majority of scientists believe the public has a right to know what is going on behind the laboratory doors.

"Since World War II there has been a group of serious science writers who are knowledgeable. They've changed the course and relationship of the science community with the outside world for the better. With the recent cutback by the government of funds for scientific projects there has been a definite change of heart, a kind of soul-searching by scientists. They feel now that they must cultivate the public's good will. Many also feel that the real future of science will have to be pointed in the direction of solving social problems, not ivory tower ones."

He said he faces a constant battle with scientists over his handling of stories, because "they derive an exact group of facts and they've worked very hard to be exact. They want it stated exactly, if necessary to five decimal points. On the other hand, we must adapt the item to the general public, which just isn't interested in five decimal points."

"My feeling is that if you move the public in the general direction of science you've accomplished something. Our stories are not geared to the scientist himself. They are for the layman. The readership of our paper just doesn't appreciate the intricacies of high-energy physics. As a result, about 70 percent of our science material is biologically, medically or psychologically oriented."

Julie McClure, former science editor, Atlanta Journal: She found some scientists reticent to talk and fearful of misinterpretation and of appearing like idiots to their peers. The scientist, she said, does not mind too much what the public thinks, but he must have the respect of his colleagues.

Although she was the official science editor, she said the city desk regarded her as a general assignment reporter first. She said the desk wanted primarily medical and hospital news and they "wanted it written in a sensational manner, with the gimmick angle. They don't need a writer who knows anything about science for the kind of science writing they want."

Anonymous science writer: "The world has passed by the old-time city editor. Today's youth does new math and what city editor can even explain new math? The city editor has a conception of the average reader who must be hit hard to get his attention. But this 'typical reader' will not exist in another few years, if he exists now. The rapidly climbing educational level of Americans has all but done away with this average reader many city editors cling to."

III. Science and Technology in The Emerging Nations

ABDUS SALAM

RIGHT FROM THE days of Merlin at the Court of King Arthur, the
scientist has enjoyed the dubious repute of a wizard. One of the
most famous scientist-wizards of the Middle Ages was Michael, the
Scot who was celebrated in verse by his countryman Sir Walter Scott
in the "Lay of the Last Minstrel." A traveller to the Paynim countries of the East tells us that:

> "In those far climes it was my lot
> To meet the wondrous Michael Scott;
> A wizard of such dreaded fame,
> That when in Salamanca's cave,
> He lifted his magic wand to wave,
> The bells would ring in Notre Dame!"

We are also told Michael's words could cleave the Eildon hills in
three; he could bridle River Tweed with a curb of stone; at a
sign from him you could be transported from Portugal to Spain in
the space of less than a night.

We do not know if Michael the Scot did really command the
powers ascribed to him. Even if he did, he could only have anticipated the men of Alamogordo and Cape Kennedy by just a few
centuries. We may, however, with reason inquire why he did acquire in the Middle Ages the dread reputation that haunted his
memory.

Michael the Scot was a humble scholar. Born in 1175, he was one
of those few inquiring men who wished to pursue science with
teachers who were currently creating it. At the age of twenty-five he

228

travelled to the Islamic University of Toledo in Spain; to study he had to learn Arabic, the then language of science. From Toledo he proceeded to Padua and Rome, teaching and translating what he had learned. His was the first translation of Aristotle's *De Anima* into Latin, not from the original Greek which Michael knew not, but from Arabic. His repute for wisdom, for wizardry, was a tribute, if you wish, to Arabic mathematics, Arabic astronomy, Arabic medicine of that day.

I have thus chosen to preface my account of science and technology in the developing world today with an account of Michael the Scot. The history of science, like the history of all civilization, has gone through cycles. Some seven centuries back, at least some of the developing countries of today were in the forefront of scientific endeavour; they were the standard bearers, the pioneers. George Sarton in his monumental five-volume *History of Science* chose to divide his story of achievement in sciences into ages, each age lasting half a century. With each half century he associated one central figure; thus 500 B.C.-450 B.C. Sarton calls the Age of Hippocrates; 450 B.C. to 400 B.C. is the Age of Plato. This is followed by the half centuries of Aristotle, of Euclid, then of Archimedes and so on. From 650 A.D.-700 A.D. is the half century of the last Chinese scientist I-ching. From 750 A.D. in an unbroken succession for 300 years, are the Ages of Jabir, Khwarizmi, Razi, Masudi, Wafa, Biruni and Omar Khayyam—Arabs, Turks, Afghans and Persians—men belonging to the culture of Islam. Around 1100 A.D. appear the first Western names, but the honors are still shared between the East and the West for two hundred years more. From 1350 A.D., however, science was created only in the West. No wonder then that Michael the Scot, in 1200 A.D., had to travel to Toledo to complete his scientific education. No wonder that this association with the infidel earned him an excommunication. No wonder that Dante consigned him to Hell.

I hope I shall not be accused of parochialism in reminding you thus that in the march of science and civilization other cultures, other lands, have played their humble role. This central fact is important to the theme I wish to unfold; it determines a whole set of attitudes, the whole approach of the emerging countries to acquiring modern scientific and technological competence is conditioned by it.

Now, throughout the ages its scientific and technical knowledge have influenced the material prosperity of a civilization. Technical advances in agriculture, in manufacturing methods, in transport have occurred in all human societies and these have always led to increased prosperity. But it is a central fact of human history that something unique occurred with the 18th and 19th century breakthroughs in physics and chemistry and metallurgy, something unique, something massive, something cumulative. The firm and the scientific mastery of natural law acquired in the last two hundred years gave man so much power, and led to such a great increase in production, that for the first time in man's history a purposeful application of scientific and technological techniques can transform the entire material basis of whole societies, eliminating hunger and want, and ceaseless toil and early death for the whole human race. Technical competence and material prosperity have become synonymous and it is this cardinal fact that the poorer two-thirds of humanity is beginning to realize. It is this cardinal fact that the developing world must come to grips with today.

How did this great division of humanity—the division of the rich and the poor, both materially and technologically—first come about? Clearly I have no competence to speak for all developing countries; there is one part of the world, however, I know much better than any other—my own country, Pakistan. Instead of generalities I propose to give you a detailed picture of Pakistan's technological past, its present and its hopes for the future. In many ways the problems I shall deal with are typical of the rest of the developing world. In particular I shall show you how important an impact on modern Pakistan the imported technology of the 19th century had. And I shall endeavour to show how many of its problems arise because we did not adjust to the technology of the 20th century in time.

I shall start my story about three centuries ago. Around 1660 two of the greatest monuments of modern history were erected—one in the West and one in the East; St. Paul's Cathedral in London and the Taj Mahal at Agra in the India of the Great Mughals. Between them these two monuments symbolize, perhaps better than words can describe, the comparative level of craftsmanship, the comparative level of architectural technology, of affluence and sophistication the two cultures had achieved at that epoch of history.

But at about the same time was also created, and this time only in the West, a third monument, a monument still greater in its eventual import for humanity. This was Newton's *Principia* published in 1687. Newton's work had no counterpart in the India of the Mughals. The impulse-springs of Islamic science had dried up earlier. The Taj Mahal was about the last flowering of a tradition, a tradition that was no longer creative, a tradition that was soon to wither and die.

The two cultures, the two technologies, that of the East and the West, came into sharp impact within a hundred years of the building of the Taj. In 1757, to be precise, the superior fire-power of Clive's small arms inflicted, on the battle grounds of Plassey, a humiliating military defeat on the descendants of the great Mughal. Another hundred years passed and in 1857, the last of the Mughals had been forced to relinquish the Imperial Crown of Delhi to Queen Victoria. With him there passed not only an empire, but also a whole tradition in art, in technology and in learning. By 1857, English had supplanted Persian as the language of state; the medical canons of Aricenna had been forgotten and the traditional art of fine muslin-weaving in Dacca had disappeared to give way to the cotton prints of Lancashire.

But from the decay of the Mughal state in 1857, from the embers as it were, there also arose the beginning of a new and modern state—the state of Pakistan in the northwest and east corners of the Indian sub-continent.

West Pakistan, of which I shall principally speak today, is a state twice as large in area and thrice as populous as California. It is a dry, parched land, watered by the mighty Indus and its five tributaries, the Ghelum, Chenab, Ravi, Beas and Sutlej. In 1857, when the British came to it as conquerers, they found ribbons of cultivation a few miles wide on either side of each of the five rivers. Between these cultivated ribbons lay stretches of parched desert.

Not content with these ribbon-like patches of cultivation, some far-sighted men in the Indian Civil Service harnessed the technology of their day to create a garden out of the scrub and the desert. They built low dams across the five rivers on foundations of gravel and sand—structures which had rarely been attempted elsewhere and whose essential stability remains something of a miracle to most hydraulic engineers till this day. Behind the dams they diverted the waters into great new canals. These canals have a total length of

10,000 miles; some of the biggest are as large as the Colorado River. With the canal system was created a fine railway network and perhaps the best road system east of Suez. West Pakistan, as a consequence, multiplied in prosperity, fertility and population. It was in this sense that modern Pakistan—in company with many another ex-colonial country—was a creation of the 19th century technology.

But even after such a heavy initial dose of technology, and again very typically, Pakistan failed to become a technologically advanced country. Something went grievously wrong, for the first flush of prosperity lasted no more than a few decades. The country was built upon just one resource, agriculture, and agricultural production did not keep in step with the population increase. Even on the purely technical side, soon after its inception the very miracle of West Pakistan's canal network began slowly to stifle the fertility it was meant to create by spreading the blight of waterlogging and salinity in areas through which the canals passed.

The reasons for this failure of technology were not difficult to find. The technology which created Pakistan did not touch us more than skin-deep. It was a graft that never took, not something that became an integral part of our lives, and all for one basic reason. The only way to communicate the garnered wisdom and knowledge of one generation of men to another is by precept and education. It may be hard to believe, but in Pakistan no provision whatsoever was made for scientific or technical or vocational education.

Thus, even though the entire object of bringing the canal waters was to increase agricultural production, no one dreamed of introducing agricultural technology in the educational system. Something like thirty-one liberal arts colleges were built in the country, one in every district headquarters, to teach British history, the metaphysics of Aristotle, the laws of equity and the principles of jurisprudence, but the whole of West Pakistan and the whole of East Pakistan had to be content with just one agricultural and just one engineering college—and this for a population then approaching fifty million. I cannot begin to convey my own personal sense of disbelief when some years ago I was told that in the United States every state university grew around the nucleus of an agricultural faculty with every other faculty added later, so contrary this was to anything I had been used to in Pakistan.

The story repeated itself at all levels. The results could have been foreseen. The level of agricultural practice remained as static as under the Mughals. The chemical revolution of fertilizers and pesticides touched us not. The manufacturing crafts went into complete oblivion. Even a steel plow had to be imported from England.

Why did the British administration fail to place any emphasis on technical education, on mechanical skills, or on agricultural husbandry? For mechanical skills there is perhaps a simple explanation. In the economic organization of the empire, Britain was to be the only manufacturing unit. All its other parts—like the American colonies, India, Nigeria, Sudan—were to supply raw materials. Thus, from a British administrator's point of view there was no need to foster mechanical or industrial skills, for these would never be exercised. But by the same token this attitude is harder to understand so far as agriculture is concerned; the attitude which, for example, failed to build up a proper agricultural advisory and extension service.

Perhaps there is a simpler and more charitable explanation possible. The educational system of the British India was essentially the creation of one man, the great historian Lord Macaulay. Writing his recommendations in 1835, he strove to give us the best that Britain could then offer. This best unfortunately did not embrace science and technology. In so far as Britain's industrial revolution had been brought about by gifted amateurs, there was in the Britain of the 19th century no appreciation of the role of technical education in fostering industrial growth. The first Royal Commission on Technical Education did not report until 1884. The first parliamentary grant in Britain for scientific and technical education amounted to no more than three thousand dollars. The first polytechnics did not come until 1890. Unlike Germany, modern Russia, or the United States, Britain did not build up an industrial society through the medium of education. Whether or not Britain suffered at all in the long run I shall not say, but for Pakistan, whose whole educational system was patterned on that of the British, this was disastrous.

In 1947, after ninety years of foreign rule, the nation started on a new phase of life. For us in Pakistan the struggle for independence has been fought on two fronts, one against the British for liberty and the second for recognition of our separate existence. On the debit side, we started with a desperately poor population, with a per capita income of fifteen cents a day. We started with no manufacturing capacity or skills and we started with a primitive agriculture, with one-fifth of our cultivated area bedeviled by the twin curses of salinity and waterlogging.

On the credit side, however, we had two assets. First, the revolution of rising expectations had hit Pakistan as strongly as it hit the rest of the underdeveloped world. Second, although there was no clear notion of how to effect an economic transformation of society, there was no resistance to newer ideas or to a newer organization of life. Like every nation smarting under recent defeat of arms, we too had passed through the phase when everything Western was an anathema, but in 1947 this phase of our history was a long way behind us.

We spent the first ten years of our independent existence in trying to redress almost feverishly, and perhaps with complete disregard of sound economics or personal suffering of the consumer, the imbalance of industrialization. The basic consumer industries—like textiles, sugar, cement and paper manufacture—were hastily created by private enterprise. But it may perhaps be right to date the era of our purposeful growth from 1958, the year that President Ayub Khan came into power and the State Planning Commission started to function with the fullest vigor. About then we began receiving the maximum help from our friends and allies, not the least from the United States, and since then we have saved and invested yearly some twenty per cent of our national income.

For the last three years the economy has grown at the rate of six per cent, the highest in Asia. Industrially we have reached the maturity of being able to project a modest heavy industrial and chemical complex based mainly on the major industrial raw material which we possess in plenty, natural gas. A modern shipbuilding yard now exists in Karachi; three refineries and two steel plants of half-million ton capacity are being erected. Since 1950, four technical universities have functioned and a number of others are projected. As a measure of the level of craftsmanship achieved, Pakistan is at present the largest net external supplier of surgical instruments to the United Kingdom. In agriculture we are on the threshold of the chemical revolution. One of the most imaginative of recent scientific missions was the 1961 team of university scientists, hydrologists, agriculturists, and engineers assembled on President Kennedy's behest by Jerome Wiesner and led by Roger Revelle to study the salinity and waterlogging problem in West Pakistan. Waterlogging and salinity are as old as irrigation itself. It has also been known for long that proper drainage is the only answer, but what makes horizontal drainage impossible in the Indus plain is the unfortunate circumstance that the plain slopes no more than a foot per mile. Horizontal drainage would be prohibitive in cost. The Revelle team suggested vertical drainage instead—mining of fresh water from an underground reservoir known to exist by a network of deep tubewells. Some of the water would seep back underground, leaching away the salt in the process. Also the general lowering of the water table on account of the pumping would cure waterlogging.

I am dwelling on the work of this team in such detail for there is something important I wish to illustrate—the impact of high-caliber scientific minds on problems relatively old. Vertical drainage had in fact been tried in Pakistan for the last fifteen years. But the results were discouraging. The great contribution of the team was to stress that the difficulty came from using the method on too small a scale. A single well, for example, has no effect on the water-level because water seeps in from the surrounding areas as fast as it is removed. To achieve a substantial lowering of the water table one must exploit the simple fact that with increasing size of a surface, the area increases more rapidly than the perimeter—the same principle which in wartime Britain made the British decide in favor of large transatlantic convoys as affording better protection against submarines' peripheral attacks rather than convoys of tiny size. Revelle's calculations—using extensive digital and linear programming at Harvard—showed that if one dealt with areas no smaller than one million acres (roughly 40 sq. miles) the peripheral seepage would not win against the area pumping; one might then hope to eliminate waterlogging within a year or two. It is two years since Revelle's report was presented. Its results have been brilliantly confirmed in the last year, in a region west of Lahore. An earlier tube-well scheme covering 60,000 acres had no effect on the water level; a larger scheme covering 1.2 million acres has drawn down the water level by about two feet—with an increase of crop yield of 50-75%.

I said earlier I shall deal extensively with Pakistan and the impact of technology and science on its growth for two reasons; firstly because I am personally more familiar with the problems of my own country; secondly because I believe the picture of Pakistan is typical of the bulk of the developing world. Pakistan presents us the picture of an ancient civilization, not too distantly in a scientific and a cultural lead; a proud nation humiliated into military submission in the recent past. The defeat was followed by an introduction of newer technological ideas and the harnessing of its rivers for agricultural growth. The new pattern was beneficial in part, but there was no wholeness to it; there was no fostering of a whole scientific and technological tradition accompanying an importation of technology. No educational system was created to carry the mastery of newer technologies further. After the first flush of prosperity, there was the inevitable over-population, the inevitable hunger and poverty. Exactly the same pattern repeated itself in India, in China, in Indonesia, in Egypt, in North Africa and elsewhere. But what of today? The great colonial convulsions of the last twenty years have freed our nations from tutelage. We can plan and execute our own destinies purposefully, remembering the lessons of the recent past. In varying degrees we have realized that there is only one way forward; to pick up the threads of technological and scientific revolution, to bring back skills and learning from the modern Toledoes. Unfortunately the magnitude of the problems is so great, the scale so vast, that along with skills we also need large quantities of scarce capital. Science and technology are no magic wands; machines cost money to make. A Roger Revelle may make the brilliant diagnosis— area versus perimeter for Pakistan's waterlogging problem—but the tube-wells must be fabricated, not just of iron and brass but sometimes, for the very saline soils, of costly fibreglass.

The path which most developing countries are taking is more or less uniform. First and foremost is the acquiring of basic skills for the operation of a technological economy with priority on the exploitation of natural resources. These may be natural gas or oil, aluminum ores or the good agricultural earth. For the larger countries this may mean building of fertilizer-producing machinery and processing and fabrication plants. There is nothing more sensitively felt in a developing country than the feeling that those in a richer region would like to see it devote itself to primary production and no more. I said earlier the pattern all over the developing world is the same. To substantiate this let me quote to you from the United Nations Special Fund report. This fund operates with a capital of about half a billion dollars and is one of the brightest landmarks of international co-operation initiated by that much maligned organization. Let us follow alphabetically the requests to it of some of the various governments:

Afghanistan: Request for Survey of Ground Water Investigation.
Argentine: Mineral Survey in the Andes.
Bolivia: Mineral Survey of the Altiplano.
Brazil: Survey of the San Francisco River Basin and Rock-Salt Deposits.
Burma: Survey of Lead, Zinc Mining and Smelting.

* * * * *

Pakistan: Engineers' Training.

Everywhere it is the same; acquiring of skills, and exploitation of natural resources, more and more productive agriculture.

These are the major, the urgent tasks of science and technology today for the emerging countries. In fulfilling these we need all the help, all the co-operation we can get.

"New Clues in J. F. K. Assasination Photos, " by Judith Goldhaber. Reprinted from The Magnet, July, 1967, Volume 11, No. 7, published by the Lawrence Radiation Laboratory, Berkeley and Livermore, California.

Judith Goldhaber

NEW CLUES IN
J.F.K. ASSASSINATION PHOTOS

Working with a pair of draftsman's calipers and an old copy of *Life* magazine, LRL physicist Luis Alvarez has made a discovery which appears to be one of the most important pieces of technical evidence ever brought to light in connection with the assassination of President John F. Kennedy.

Alvarez' discovery, backed by independent evidence obtained in a carefully-controlled series of empirical tests, was featured in the recent CBS television four-part special report on the assassination and the Warren Report. Prior to the broadcast, few of the LRL scientist's friends and co-workers were aware that the well-known Alvarez ingenuity had been focused on the assassination.

The essence of Alvarez' contribution is the discovery that the Abraham Zapruder motion-picture films of the assassination, printed in *Life* and in the Warren Report, contain internal clues which may reveal the number of shots fired and the exact instant when they were fired. The clues are streaks on the film, which Alvarez believes could have been caused only by a sudden involuntary jog of Mr. Zapruder's camera. Alvarez suggests—and the independent tests impressively corroborate — that such a jog is precisely the response of a hand-held motion picture camera to a nearby rifle shot—a combination of the

direct shock wave from the bullet and a flinching reaction of the photographer.

If this theory is correct, the implications for the controversy surrounding the Warren Commission Report are very great. From his analysis of the 165 frames available to him, Alvarez was able to show that three and only three shots were fired (as the Commission maintained), and, furthermore, that the bullet which missed its mark was the *first* shot, fired just before the one that hit the President in the back. (The Warren Commission inclined to the belief that the *second* shot was the one that missed.) This finding is crucial, since it significantly lengthens the amount of time available to the assassin to aim and fire. Instead of the 5.6 seconds previously thought available, Alvarez' evidence points to more than seven seconds between the first and third shots. This would make the "single assassin" theory much more plausible to those who have been critical of the Warren Report's conclusions.

Alvarez' analysis of the film (later corroborated by photo analysis experts at Edgerton, Germeshausen and Grier, Inc.) indicate that the three shots were fired around frames 180, 220, and 313 of the Zapruder film. The latter two frames are the ones that were previously

238

believed to correspond to the time of the rifle shots. Frame 180, however, is a surprise, since it occurs in a period when it had been thought that no shots were fired—while the presidential limousine was passing under a tree which screened it from the sixth-floor window of the book depository. Significantly, however, frame 180 coincides exactly with the brief moment when the limousine passed through a fairly large gap in the foliage, offering for a few seconds a clear line-of-sight to the assassin. Alvarez suggested that that first bullet may have hit the tree, and may, in fact, still be lodged there. At his urging, CBS scanned the tree with a metal detector—but without success.

Empirical Tests

The independent tests of Alvarez' hypothesis were performed, at his suggestion, by experts at E.G. & G., the firm that also does much technical-photography analysis in connection with LRL Livermore field activities. In the tests, experimental subjects holding cameras identical to Zapruder's filmed a target while rifles were fired in the same relative position as the assassin and Zapruder. Even though the subjects had prior expectation of an impending disturbance, and were instructed to keep their cameras as steady as possible, every subject reacted to the shots with exactly the sort of involuntary movement that Alvarez had predicted. Furthermore, these movements produced streaks on the film closely resembling the effects found in the Zapruder photos.

Whether or not Alvarez' theory is ever confirmed conclusively by further evidence, it has already been acclaimed all over the country as a masterful piece of analysis, in the best tradition of closely-reasoned, imaginative scientific detection. What makes the accomplishment particularly impressive is the fact that Alvarez used no complicated technical apparatus and had no special access to materials or evidence unavailable to the rest of the public. He worked entirely from the *Life* magazine color photos and the black-and-white reproductions in

the 27 volumes of the Warren Commission Report—pictures which have already been scanned and re-scanned by millions of people, and which have been subjected to the most exhaustive technical analysis in FBI laboratories.

Alvarez himself attributes his success to two seemingly unrelated interests out of his scientific past: first, his experience in studying shock waves and designing instruments to measure them; and, second, his longtime interest in photography, which has led him to study the phenomenon of camera "jitter" in hand-held motion picture cameras. It is unlikely, says Alvarez, that anyone else with those two particular specialties in his background had ever happened to examine the Zapruder photos in detail up until this time.

The Wrong Clue

Like so many important discoveries, this one began with a mistake—a mistake which cost Alvarez many hours of painstaking labor, and yet which led him, finally, to the discovery of important new information.

The mistake involved a flag—specifically, the American flag on the right fender of the President's limousine.

As Alvarez recalls it now, he first became interested in the challenge of applying scientific analysis to the assassination photos on the day before Thanksgiving, 1966, after a lunchtime conversation in the LRL Berkeley cafeteria with some of his graduate students, who were arguing heatedly about the Warren Report's conclusions. That evening at home, he pulled out the copy of *Life* with the Zapruder photos and sat down to examine them carefully for the first time.

Almost immediately, his attention was caught by the appearance of the flags on the President's car. The right-hand flag seemed to be different in frame 228 than in all other frames. Is it possible, he wondered, that the flag was reacting to a sudden shock wave? And could that shock wave have been caused by one of the bullets? With growing excitement, he measured the flag's apparent width in successive frames, applied elementary formulas to calculate the acceleration of the edge of the flag, and convinced himself that the flag might indeed be reacting to some unusual stimulus. That conviction turned out to be wrong. The flag distortion, Alvarez now feels, was due to wind rippling, and can be found in several frames not conceivably associated with rifle shots. But that realization did not come for several more days —and by that time, Alvarez was hooked.

Alvarez was now anxious to check his rough calculations with other photographs contained in the 27 volumes of the Warren Commission Report. But it was Thanksgiving weekend, and the University libraries were closed. Too impatient to wait, he confided his ideas to a friend, Edwin Huddleson, a San Francisco lawyer, who told him about a library that would be open—the law library in San Francisco's City Hall.

All through the remainder of that Thanksgiving vacation, Alvarez worked in the law library, measuring and re-measuring, calculating and re-calculating, taking voluminous notes. By the end of the vacation period, the initial "flag" hypothesis was all but forgotten, and Alvarez was on the track of something much more exciting.

Frame 227

The important breakthrough came when Alvarez noticed that frame 227 of the Zapruder film shows the highlights on the car's windshield spread out into streaks about 2 millimeters long, oriented about $25°$ with the horizontal. Further examination revealed that many other objects that had been in sharp focus in previous frames were streaked or blurred in this frame.

Alvarez' notes, which he made available to the MAGNET for this article, furnish a vivid picture of the slow process of discovery and the awakening of insight:

Monday, November 28: "Note streaked highlights in frame 227 . . .

Tuesday, November 29: "227 remains the most puzzling picture . . . The extra-

ordinary thing is that neither the men in the right middle or the squares in the background seem to be at all smeared . . ."

Soon after the latter entry, the insight came. Here's how he described it in a letter he wrote to Huddleson a few days later:

". . . even after I had found the streaking in 227, it took me some time to realize why everything on the north side of the Elm Street curb was streaked, while the objects on the south side of the curb were unstreaked — the background squares, the tree, the two men . . .

"I finally saw that the man waving in the foreground wasn't streaked—the crease in his shirt at the elbow shows that he and the camera weren't in relative motion. So finally, in a most tedious way, it became clear that what had happened was simply this: the camera axis, which had been panning smoothly to keep lined up on the moving car, had been given a sudden twist to the left (counterclockwise). This completely stopped its panning motion, so that both the foreground man and all the background objects became sharp. But the fact that the car and the motorcycle policemen were moving to the right produced the drawn-out streaks of light on the film."

What could have produced this sudden uncontrollable twist to the left in such a good photographer as Mr. Zapruder, who displayed a high level of skill and coolness in all other aspects of the filming? Were there other instances? Alvarez quickly searched all the other frames for similar streaks—and found them in several places. He then plotted the horizontal spread of the highlights in every single one of the 165 frames reproduced in the Warren Report, and found three streaking "peaks." The first group of streaks starts at frame 182 and ends at 202—just over one second of time. The second group is the one discussed above, in connection with frame 227. In frame 313, the most clearly defined shot hit the President. Streaking shows the camera jerked violently to the right at frame 313. Alvarez believes that this frame registers the direct effect of the bullet's shock wave—the same effect that breaks windows during sonic booms. In the other streaked frames, he believes that the greater part of the camera's motion was caused by Zapruder's neuromuscular reaction to the sound of the rifle shot.

Testing the Theory

By December 4, Alvarez was sufficiently convinced of the correctness and importance of his theory to communicate it to someone else. But, before going too far out on a limb, he wanted to examine the original Zapruder film and have independent tests of flinching reactions performed at a well-equipped laboratory. He thought of Frank Stanton, president of CBS, chairman of the Rand Corporation board, and a longtime acquaintance—someone who would certainly be in a position to cut through red tape and arrange the necessary clearances. Therefore he summarized his conclusions in a 30-page typewritten letter and sent it off to Huddleson for transmittal to Stanton. The CBS president's reaction was quick and enthusiastic. Within a few weeks, the E.G. & G. tests were planned and executed. On January 19, Alvarez met with CBS and E.G. & G. officials in the National Archives in Washington and studied the Zapruder originals, which were consistent with all of the ideas suggested by the 165 frames printed in *Life*. (The E.G. & G. experts analyzed this film for evidence of additional shots, and found none.)

Alvarez' conclusions from his analysis of the Zapruder film are contained in two letters—the 30-page one mentioned above and a follow-up letter dated January 2. The letters discuss several different aspects of the film—not all of which were covered in the CBS report. Among the material not followed up in detail by CBS is a discovery (here described publicly for the first time) that Alvarez himself considers to be of considerable importance. It relates to the question of whether Zapruder's Bell & Howell camera, which was supposed to be operating

at a speed of 18.3 frames per second, was actually running at that rate during the filming. The camera's speed is critically important, since the Zapruder film has been used as the basis for all reconstructions of the assassination; if the camera was running fast, the time between shots would be correspondingly shorter. Many experts have addressed themselves to this problem, but it has generally been regarded as essentially insoluble. Of course, the camera was itself tested later, but such tests do not necessarily prove that it was running at its rated speed *at the time* when the critical photos were taken. For example, some critics have suggested that Zapruder might, in his excitement, have depressed the camera's button to the high-speed position, thus increasing the frame rate to 1.5 times the original rate.

An Internal Clock

Alvarez believes that he has found an internal "clock" in the film that proves conclusively that the frame rate was *not* significantly fast compared to the 18.3 frames-per-second estimated by the FBI. The clock is easily visible and—after it has been pointed out—almost absurdly simple. In frames 278-296 of the Zapruder film, there is a man in a white shirt, standing in front of a small boy and a woman. The man is clapping. His hands are together in frames 280, 285, 291, and 296, and are farthest apart in frames 278, 283, and between 287 and 288. From these pictures, it is a simple matter to figure out the frequency of the man's clapping. If the camera is running at 18.3 frames per second, the man is clapping at four claps per second; if the camera is running 50% faster (at 27 frames per second), the man is clapping at 6 claps per second. If the camera is running twice as fast (at 36.6 frames per second), the man is clapping at eight claps per second. Yet it is virtually impossible for a human being to clap at either of the two higher rates. This sounds implausible, but it's true: if you wish, you can test it yourself exactly as Alvarez did, by synchronizing your clapping to a metronome. The reason for this odd circumstance is the fact that the muscular power a person must exert to clap (with a constant hand spread) varies as the cube of the clapping frequency. To double the clapping rate, you must increase the power expended eight times. A reasonable clapping rate (even for a very enthusiastic spectator) is four or possibly five claps per second; six and above is unreasonable. Thus, by internal evidence, it appears that the frame rate of Zapruder's camera can be closely restricted to 18.3 frames per second plus about 15-20% *at the most*. This variation is not large enough to significantly affect the time-reconstruction of events based on the film.

Alvarez believes that there are probably still many more clues in the Zapruder film and other photos of the assassination that could be brought to light by careful photo analysis. For example, it has generally been believed that it is impossible to pinpoint the exact position of the presidential limousine in each frame of the film, because of the essentially "featureless" background. Alvarez, however, has found several tiny "features" which he believes could be used to determine the position of the car from the beginning to the end of the critical period.

Back to Alamagordo

As for himself, Alvarez intends to pursue the analysis of the films no further. Nor is he tempted to become involved in other aspects of the investigation—even to the extent of reading any of the mountain of books that have been published on the subject. He looks at his accomplishment, however, as an exhilarating exercise in the principles of scientific inquiry—not too different, in essence, from the other puzzles in logic that have intrigued him during his long career. In fact, he ends his summary of the investigation with a reminder of an "earlier use of shock wave-induced motions of lightweight objects"—one that should be very familiar to LRL readers. At the Alamagordo test of the first atomic bomb, hundreds of scientists were manning the most expensive and sophisticated equipment available for measuring the overpressure in the shock wave, which would give the equivalent kilotonnage of the explosion. One man, however, had a simpler method. As Lansing Lamont described it in his book *Day of Trinity*, "Enrico Fermi didn't notice the crack of the shock wave . . . [He] was too engrossed in dribbling scraps of paper from his pockets. He watched them slowly fall, then sweep across the reservoir as the shock wave struck them. Within seconds, Fermi had paced the distance the scraps had blown and estimated the force of the explosion as the equivalent of 20,000 tons of TNT."

Fermi's measurement, Alvarez notes, has stood the test of time.

RADIO ECHOES
(THE ORIGIN OF RADAR)

Excerpt from an address at Girard College, Philadelphia, on receiving the John Scott Award, December 15, 1948.

*Merle A. Tuve**

Before I close this talk, I think I should tell you a little story, which illustrates how innocent of great things a scientific man feels at the time of a discovery. I was privileged 23 years ago to participate in the first discovery of the echoes of radio waves from the ionosphere, which is the electrically conducting region 100 miles or so overhead which reflects radio waves around the world and makes communication possible, instead of letting the waves fly out into space. Our radio echoes developed into something unexpected a good many years later.

Dr. Gregory Breit, now a Professor at Yale, was a new member of our Department of Terrestrial Magnetism in the autumn of 1924, when I was a young instructor at Johns Hopkins in Baltimore. In 1885 an electrically conducting layer in the upper atmosphere had been postulated to explain the daily variations of the compass. Marconi sent his famous letter "S" across the Atlantic in 1901 and the same conducting layer was the only explanation for receiving his radio waves in Newfoundland around the curvature of the earth. Dr. Breit wanted to observe the reflections of short radio waves from this layer, to be sent upward from a big parabolic reflector in

243

Washington, and he asked me to devise the receiving arrangements in Baltimore, 40 miles away.

A meeting of various experts was held at the Department, the problem was considered, and money was approved for a reflector 100 feet high. I waited until the money was in hand, and then confessed to them that I was worried about the difficulties of receiving such short waves (under one meter) at such a distance as Baltimore, after an uncertain degree of reflection by the upper atmosphere. I had previously found it difficult to receive them at a distance of 30 feet.

On the spur of the moment I suggested to the meeting an old and familiar idea, namely, that it might be better to try for echoes instead by sending out short radio pulses, because in this case longer radio waves could be used. Incidentally, my worries were correct; we know now that waves short enough for use with the reflector would have gone right on through the upper atmosphere. At dinner that evening Dr. Breit and I planned the apparatus and saw that the echo idea might explain some other things we knew about, such as poor modulation and rough music at 150 miles from a station which sounded all right when you listened nearby.

Two months after this meeting Professor Appleton in England published a proposal for measuring the height of the ionized layer by slowly varying the frequency of a continuous wave, which experiment he carried out in 1925. Meanwhile, during the winter and in the spring of 1925, Dr. Breit tried some "echo" experiments using broad pulses ("dots") on distant transmitters, observing the received signals on a cathode-ray oscillograph and recording with a Duddell oscillograph. The received signals showed extra peaks, which might have been echoes, but there was no way to be sure that these were not from rough pulses at the transmitter, due to sparks at the keying device.

I came to Washington to spend the summer of 1925 at the Department working with Dr. Breit on this echo problem.

We had decided that the only way to be certain about the echoes was to use vertical reflections, with the transmitter adjacent to the receiver so that we could be sure by oscillographic observation that the transmitted pulses were clean while we simultaneously observed the extra pulses, due to echoes, on the receiver.

We persuaded Dr. A. H. Taylor at the Naval Research Laboratory to let us apply pulse modulation to a Navy transmitter, and we installed our receiver and oscillographs there.

On July 10, 1925, we first observed echoes, using 80-meter waves. They were very clear, and the transmitted pulses were without extra pulses. However, the distance indicated by the echoes corresponded to the distance to the Blue Ridge Mountains, and it remained constant. This continued for many days. We were not after echoes from distant objects, we wanted to study the upper atmosphere, and we felt disappointed. Ten days later, however, we were able to use the transmitter during the evening hours, and we saw that the distance had changed; in fact it changed still more while we observed. We then knew that at least some of our echoes were from the upper atmosphere.

With our several colleagues we continued to observe echoes and improve the experiments during 1926 to 1928. We made pulses with a duration of 2/10,000 second and separated the multiple echoes. We worked at various shorter wavelengths.

During one of our experiments, which we called the "echo-interference experiment," we were greatly troubled by planes taking off and landing at an airport two miles away. We were not interested in airplanes, and had to wait for the air to clear. We also saw various other transient echoes and effects which we were not interested in. We talked freely about our experiments, and published the echo measurements. Experiments with very short waves, with movable parabolic reflectors, were considered, too, but this was not feasible because no vacuum tubes were then available for transmitters at such short wavelengths. In 1929 we persuaded the Bureau of Standards to adopt the echo method, and we stopped our experiments for three years.

The idea of using radio waves reflected from planes and ships as a military device was privately recognized and made a Navy secret by Lieutenant W. S. Parsons at the Naval Research Laboratory in 1932, seven years after our first experiments there. Today he is Admiral Parsons, the atomic bomb expert.

Professor Appleton and others in England adopted our pulse technique for upper-atmosphere studies in 1929, and carried it forward with results which surpassed our own studies. In 1934, five years later, the British military services independently saw the values of pulse-radio echoes from planes and ships, and quietly made the idea their secret. When the United States and Britain exchanged secrets in 1939 or 1940, both sides were then surprised. At about that time, roughly 15 years after our first echo observations and our disappointment and concern about echoes from the Blue Ridge Mountains, a British physicist devised a vacuum tube which would generate pulses of high power from very short wave transmitters, and the use of movable parabolic reflectors became feasible.

Radar, a simple application of pulse-radio technique and the observation of timed echoes, was a slow but direct outgrowth of our experiments on the upper atmosphere, and was probably the most important single technical device used in World War II. The patent lawyers of RCA cited our early published experiments in preventing the patent offices of various governments from granting anybody a basic patent on radar. Dr. Breit and I were naturally pleased by all these developments, and glad that our part in it was a gift to the public from research activities in pure science.

Chapter Six

ENERGY AND MAN

To open this chapter, Robert E. Neil presents an historian's view of the population explosion. He points out the key role played by the productive energy available to man. This energy increased dramatically with James Watt's invention of the steam engine near the end of the eighteenth century, thus ushering in the industrial revolution. Man now had much more than enough energy available to maintain his metabolism and to raise more food; he could improve his standard of living on a massive scale.

Yet, to stay alive and be active, each individual needs both food and oxygen, since it is the interaction of these two components that releases energy for body warmth and activity. Unfortunately, the nonuniform distribution of food leaves many on earth at the edge of starvation, unable to advance above bare subsistence. This fact is portrayed poignantly by Moritz Thomsen. And, although air and the oxygen it contains appear to be freely available, man's ability to utilize them is bounded. Roger Bannister describes how the body's need to process oxygen influences the training and performance of athletes. Closer to most of you is the chart prepared by Justus J. Schifferes, where you can find the total energy expenditure of your body when engaged in various activities, assuming that sufficient food and oxygen are available.

From the energy requirements of man himself, the chapter turns to one of man's uses of energy to improve his living conditions. Most dramatic of these in the industrialized nations has been transportation by means of the automobile. In the United States, for instance, the daily per capita energy consumption by automobiles, trucks, and buses is 30,000 Calories, more than ten times the bodily energy needs of one person. It is also more than all other energy uses for industrial and residential purposes combined! Yet the automobile is under attack because of its chemical waste products, which pollute the air we must breathe. Jay A. Bolt nevertheless expresses his faith in the gasoline-powered internal combustion engine as the prime source of motive power. Indeed, it seems impossible to do as well by any other means.

In view of the role played by energy in modern society, one might propose that the wealth of nations be measured by their energy sources rather than by land or human labor. After all, the value of land for crops and labor for production may soon be completely superseded by synthetic nutrients and

248

automated factories. Or is something overlooked in this theory? Where does the wealth of nations lie?

About the Contributors

ROBERT E. NEIL (1931-   ) is Associate Professor of History at Oberlin College. Though his professional area is German history, he has been studying problems in demography and population growth as a side interest.

MORITZ THOMSEN (1917-   ) has now settled in Ecuador. He grew up and was educated in the United States. After twenty years of farming in Washington and California, he joined the peace corps and served in a small rural community in Ecuador. Moritz Thomsen is a pen name.

ROGER G. BANNISTER (1929-   ) is a physician in London, with special interests in the physiology of exercise. He was the first athlete to run one mile in less than four minutes, holding the world record of 3:59.4 minutes for awhile in 1954 while he was a medical student. Since then he has researched and written about the limiting factors in human physical performance and has advised in relation to athletic competition in the Olympic Games and international meetings.

JUSTUS J. SCHIFFERES (1907-   ) is an American author and lecturer in the fields of health, nutrition, and physical fitness. He has served with government agencies and at universities. Mr. Schifferes is also interested in dramatic arts and has written several pieces for the stage.

JAY A. BOLT (1911-   ) is Professor of Mechanical Engineering at the University of Michigan. He is responsible for much of the research and teaching related to the internal combustion engine and other vehicle power plants at the university.

The Mushroom Crowd

By ROBERT E. NEIL, '53

WHEN I WAS TRYING to decide last February what I would talk about this morning, I was advised by one of my more experienced colleagues that there are only two real possibilities for a Senior Assembly: you can talk about the state of the College, or the state of the world. Since I couldn't make up my mind between those two alternatives before the deadline for handing in a title, I submitted an ambiguous one — "the mushroom crowd." I figured that with a title like "the mushroom crowd," you could talk about either Oberlin or the world. But I finally decided that the austerities that certainly face mankind are even more important than those which are allegedly conducive to learning at Oberlin, and so I am going to talk about the population explosion.

I realize, of course, that you've already had one talk this year on the population explosion. But since this is undoubtedly the most important long-range problem that faces us — and by "us" I mean the entire human race — I don't think that it's excessive to spend one hour a year thinking about it. And since I am an historian, perhaps the most appropriate way for me to begin is by trying to set the population problem in its historical context.

Technologies and Population Ceilings

If we stand back and take an over-view of the entire span of human history, it is clear that man's story breaks down into three distinct stages — and I don't mean "ancient, medieval, and modern." What sets these three stages off from each other is the level of technology that man has achieved in each of them. Over the millennia, in other words, man has devised three basically different ways of making a living, three different solutions to the ever-present economic problem of subsistence. Each of these three technologies has built into it certain implications for human demography. For one thing, each of them establishes a theoretical maximum for the human population. That is, for a given mode of life there is a definite limit to the so-called "carrying capacity" of the earth with respect to man's numbers. This theoretical maximum or carrying capacity is often referred to as the "population ceiling" of the technology in question. In addition to having a built-in population ceiling, each of the three technologies with which man has experimented has been associated historically with a characteristic pattern of population growth. The way man makes his living thus not only determines the theoretical maximum on his numbers but also — historically at least — sets the rate at which he is likely to multiply towards this limit.

The first stage of man's development comprises the nearly 600 millennia of the Old Stone Age or paleolithic period. Paleolithic man had a very primitive technology: he was merely a food-gatherer. That is, he lived solely by hunting and scrounging. One authority has described this primitive mode of life very nicely as follows: "For thousands and thousands of years — in fact for the greater part of his history — homo sapiens remained incapable of doing anything better than dashing all over the place trying to capture or to collect any edible plant or animal in sight. His knowledge was basically limited to what was edible and what was not." And, of course, sometimes he made mistakes — with fatal results.

Since it takes two square miles per person to support a hunting economy, and since the number of square miles on the globe suitable for hunting was limited, paleolithic man had a low population ceiling — an estimated theoretical maximum of no more than 20 million. Moreover, given the precarious nature of the hunting life, he must have had to breed at a rate close to the biological maximum merely to avoid becoming extinct. Accordingly, his population growth rate was also very low — an estimated .02% per millennium. (Paleolithic man would have appreciated Mark Twain's story about the stagecoach travelers who kept riding hour after hour trying to get to a little town in Nevada. But every time they stopped to ask how far it was, the answer was always "ten miles." Finally one passenger remarked philosophically: "Thank God we're holding our own!") Indeed, the paleolithic growth rate was so slow that by the end of the Old Stone Age the actual number of humans probably did not exceed 5 or 10 million. Man, in other words, spent the first 99% of his existence building up a population no larger than that of the state of Ohio or perhaps the city of New York today. Until comparatively recently man was a very rare animal.

He moved considerably faster in his next technological stage, however. In the Agricultural Revolution, beginning around 6,000 B.C., he changed over from food gathering to food producing, from hunting to farming. Since this was a much more productive economy, its introduction at once raised man's population ceiling to about 1 billion. This is the theoretical carrying capacity of the earth for a purely agricultural economy — an economy, that is, where farming is carried on without the help of such things as tractors and artificial fertilizers which only an industrial economy can provide to its agrarian sector. Moreover, by providing man with a better food supply the Agricultural Revolution reduced his death rate with respect to his birth rate and thus gave him a higher rate of population growth as well as a higher population ceiling. The growth rate now rose from .02% per millennium to .06% per year. This sounds like a modest increase, but it was enough to cause man's numbers to mushroom. Having spent the first 99% of his existence getting up to at most 10 million, in only the next eight millennia, before the beginning of the Industrial Revolution around 1750 A.D., man suddenly shot all the way up to 700 million. To make this sudden increase more graphic, suppose that we put man's entire existence on the time-scale of a single day: it took man 23 hours and 41 minutes to make his first 10 million, but only 18 minutes after that he was up to 700 million. (Well, as Joe Kennedy is supposed to have remarked: "The first 10 million are always the hahdest.") From this it is clear that the population explosion that we read about in the papers is actually the *second* such upheaval that man has experienced. In both cases a technological revolution, by disrupting the equilibrium between mortality and natality, has increased the rate of population growth both in absolute terms and with respect to the population ceiling.

With this background in mind, we can state the essence of today's population problem as follows: man's third technological experiment, industrialism, has generated a rate of population growth incompatible with the population ceiling of this technology. At the two previous economic levels, though the human population was constantly growing, the rate of multiplication towards the theoretical maximum was so slow that there was little practical danger of population saturation. At our present level of technology, however, and at our present rate of population growth, we will hit the ceiling in only a century and a half — a point no more remote in time than Napoleon's vacation at St. Helena. Should this actually occur, there is good reason to believe that our industrial technology would not be able to survive the overpopulation that it had produced. Mankind would then be forced to revert to a more primitive economy with a much lower population ceiling. And in that process the excess warm bodies would simply have to grow cold.

To document this, let us have a closer look at the Industrial Revolution, which began in England in the second half of the 18th century, spread to the rest of Europe and to North America in the 19th, and is now, in the 20th, struggling to take root in Asia, Africa, and Latin America.

Productivity versus Reproductivity

If we leave out of account all of the superficial trivia about flying shuttles and spinning jennies so beloved cf textbook writers, the essence of the Industrial Revolution is this: where in the Agricultural Revolution man made the transition from food gathering to food producing, in the Industrial Revolution he has changed over from ani-mate to inanimate power. Pre-industrial man was always poor, and he was poor because he lacked adequate power. As long as he had to rely for power mainly on his own muscles, or his wife's, or those of other animals, man was chronically short of productive energy — and this is one reason, by the way, why slavery and serfdom are so com-mon in agrarian societies. When, however, he learned how to harness inanimate sources of energy — first fossil fuels, now nuclear fission and fusion, and in the future most likely solar and tidal power — then *homo sapiens* at last had a supply of energy at his disposal sufficient to raise his productivity above the subsistence level for the first time. It was only then, for instance, that it became possible to exempt more than 10% of the population from the business of simply producing food. Prior to the Indus-trial Revolution, in other words, an urban civilization in our sense of the word was impossible, since it took nine men working on the land to support only one man working in the town.

The trouble is that industrial man has raised not only his productivity, but also his reproductivity, and the latter is now threatening to outrun the former. This is the result of an unusual combination of circumstances. Like the Agricultural Revolution, the industrial one has raised the world population ceiling dramatically owing to the enor-mous increase in per capita productivity. And the Indus-trial Revolution, again like its predecessor, has also accel-erated population growth by further lowering the death rate with respect to the birth rate. But this disruption of the equilibrium between mortality and natality has been accomplished by different means in different parts of the world; and this is the cause of the worst — though by no means all — of our current population problems.

In today's advanced countries, the reduction in mortality was accomplished by two factors working together and coming gradually: on the one hand, the raising of the living standard through industrialization, and, on the other hand, the improvement of the health standard through the development of modern medicine. Because these two factors were introduced gradually in today's advanced countries, the death rate was pushed down step by step instead of all at once, with the result that the acceleration of population growth was spread over a considerable period. Since this was the case, and since living standards were rising at the same time that the death rate was falling, it was possible in today's advanced countries to bring about a parallel though somewhat delayed reduction in the birth rate before the population had mushroomed out of sight. Modern medicine made it possible to control births as well as deaths, and the rising living standard made people *want* to do this. In the last two centuries, thus, the advanced countries went through a complete demographic transition from the agrarian pattern of high birth and death rates to the industrial pattern of low ones. Nevertheless, because birth control lagged behind death control during this transition, the population of the advanced countries grew from 145 million in 1750 to 745 million in 1950 before something approaching an equilibrium was reestablished.

Small wonder, then, that the population is running completely out of control in the backward countries. What has happened since the Second World War in countries like India is as follows: modern medicine has been intro-duced all at once instead of being developed gradually, and it has been introduced by itself from outside instead of in conjunction with rising living standards generated internally. The result is that the backward countries have experienced in the last two decades the same reduction in the death rate which it took the advanced countries two centuries to achieve. The most extreme example is prob-ably that of Ceylon, where the island's death rate was cut almost in half in a single year simply by spraying the malarial jungles with DDT. But since the reduction in mortality in the backward countries has not been accom-panied by rising living standards, as was the case in Europe, there has been little cultural pressure or mass incentive to reduce the birth rate proportionally. So what we have today in the backward countries is this: industrial death rates artificially combined with agrarian birth rates.

An Abnormal Period of History

This is an explosive combination. It is explosive because

it generates an average world population growth rate of 2% per year instead of only .02% per millennium, as it was in Stone Age times. To get some idea of what this growth rate means, consider the following admittedly propagandistic examples. If there had been only one dozen human beings — suitably sorted for sex — on earth at the opening of the Christian era, and if they and their descendants had multiplied at a rate of 2% per year, there would be 300 million people today for each individual actually alive. And if our present population of about 3 billion were to continue growing at the present rate, there would be one person per square meter on our planet's surface, sea area included, in only seven hundred years. From this statistic one demographer concludes, in his cautious social-sciency way, "it seems evident that such a rate of population increase as now prevails cannot continue."

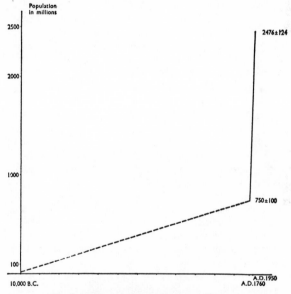

Fig. 9. **The growth of world human population,**
10,000 B.C. to A.D. 1950

Courtesy Carlo Cipolla, Economic History of World Population (Pelican Books)

Let me emphasize in this connection a point which is usually overlooked in writing about the population explosion. We tend to blame the whole thing on the non-Western world and to claim that our own growth rate has been "stabilized" at about 1%, taking the advanced nations as a group. Such things as the post-war American baby boom are dismissed, in the lingo of one writer, as "not a trend but a trendlet" — though he does not make it overly clear just when a "trendlet" becomes a trend. Well, the Chamber of Commerce types who exult in a 1% growth rate in advanced countries as "demand-stimulating" should consider the following: if an initial population of a mere 100 people had multiplied at "only" 1% per year over the 5,000 years of recorded history, there would now be 2.7 billion people per square foot of the earth's land surface. That ought to stimulate a very considerable demand!

It is clear, thus, that in contrasting the population growth rates of the advanced and the backward countries, we are not really, as some people would lead you to believe, comparing what is desirable with what is undesirable, the normal with the abnormal; instead, we are merely comparing two different degrees of intolerability — that which is intolerable right now and that which will soon become so. In the short run, we simply cannot go on, as we are at present, adding a Youngstown a day, a Cleveland a week, and an Italy a year to the world's population. And in the long run it is equally impossible to go on reproducing even at the rate now prevalent in the advanced countries that have supposedly "solved" their population problem.

I emphasize this because most of the accounts that one reads of the population explosion are actually nothing but discussions of the chances of lowering the reproductivity of the backward countries to the level of the industrialized ones. But this misses the real point. It is not a question of simply reducing the growth rates of Indonesia or Brazil to those of the United States or West Germany; instead, it is a question of reducing the growth rates of *all* countries to levels far below any that now generally prevail. We should all hope, of course, that something is done soon about the Indonesias and Brazils, but we should also realize that this would not by itself solve the human population problem; it would only constitute a first emergency step. The plain fact is that for the past two centuries man — first Western man and now the whole human species — has been going through a "fantastically abnormal period of history, and one that must inevitably end soon," to quote Sir Charles G. Darwin. A closer equilibrium between mortality and natality will have to be re-established all over the world.

The Potential Catastrophe

The easiest way to accomplish this, of course, is simply to wait and let nature do the job for us by raising the death rate as the population approaches the saturation point. The highest estimate ever made by a responsible scholar, the geochemist Harrison Brown of CalTech, of the population ceiling for an industrial economy is 50 billion. This estimate assumes that we will be able to industrialize all the countries of the world — a very doubt-ful assumption, given the way things are going in the backward countries today — and that we will be able to develop nuclear and solar power on an adequate scale to extract everything that we need from air, water, and ordinary rock. Brown's estimate also assumes

(a) That our food would be supplied entirely by "algae farms and yeast factories,"

(b) That we would be living almost exclusively in an artificial urban environment — one vast world megalopolis, and

(c) That some reduction in our present level of con-sumption would probably be required in order to develop the power sources and substitute products that would be needed to keep this human ant hill going.

Well, at the rate we are going now, we shall reach this theoretical maximum in only a century and a half. If by that time we have not stabilized our population, an almost unthinkable catastrophe will ensue. Here is how one authority thinks that it would look. A completely run-away world population would never be able to accumu-late the amount of capital needed to make nuclear and solar power available in time and in sufficient quantities to satisfy the needs of more than 50 billion people, once our supply of fossil fuels has given out. In short, man would suddenly find himself without adequate supplies of energy to keep an industrial economy going — perhaps not at all, certainly not all over the world. As a result, a large part or possibly all of mankind would then have to revert to an agrarian mode of life. But since the estimated population ceiling for a purely agrarian economy is only 1 billion, obviously the majority of the human race would soon have to perish. And the net result of our brief fling with industrialism would have been nothing more in the end than to have brought us quickly instead of slowly to the level of population saturation for an agrarian economy. The whole world would become one vast India, where already today one out of every fifteen people is kept alive only by grain imported from the United States and where in the city of Calcutta 600,000 people live and die in the streets. Moreover, if such a condition ever became estab-lished on a world scale, it would likely prove to be irrever-sible. To rebuild the collapsed industrial economy, man would need enormous quantities of inanimate power. But the cheap sources of such energy would already have been consumed, and man would be simply too poor to afford the more expensive kinds. The only "hope," if one can call it that, would lie in the forced accumulation of capital

257

by dictators whose methods would necessarily make Stalin's seem positively benevolent.

So I think that on the whole it would probably be best not to wait for nature to do the job. The question then becomes whether man will be able to do it for himself. As yet, nobody can give a definite answer to this question, but two things can already be said about it with some confidence.

What We Must Do

The first is that we know perfectly well what a long-range solution to the problem will have to include. In the backward countries we must introduce those factors which have led to a reduction in the birth rate in the advanced countries — namely, medical means of birth control and a higher living standard. With regard to birth control, a considerable amount of practical experience already suggests that mere contraceptive devices, including the pill, will not prove very effective until the educational level of the backward countries has risen considerably. In the meantime, these nations will probably have to resort to government-run programs of mass sterilization, such as India is now developing, and to legalized abortion, as in Japan where there are more than a million such operations a year. With regard to raising the living standard — the other indispensable prerequisite for curbing the population increase in the underdeveloped countries — here the only real hope is outside assistance on a scale dwarfing anything that we have yet seen. If the backward countries are left to their own devices, their living standards are more likely to go down than up; and even if they do go up, this will occur far too slowly to act as an effective brake on population growth. Having, so to speak, exported our worst social disease, overpopulation, to the rest of the world in a highly virulent form, we in the advanced countries must now send in the capital to help stop the epidemic. Indeed, some experts believe that for such aid to be effective we would have to send in so much capital that we would be required to accept a certain reduction in our own living standard, just as we would to win a war. Whether or not the advanced countries will be prepared to do that is more than doubtful, especially given the huge cost of the arms and space races.

But suppose that through adequate foreign aid and domestic birth control programs we do manage to reduce the population growth of the backward countries to something approximating the current industrial level, what then? Then we must tackle the perhaps even more ticklish problem of reducing our own birth rates still further. This problem is difficult for two reasons, which the American demographers Lincoln and Alice Day state as follows: first, "the low mortality rates of today permit a rapid, sustained population growth when average family size is of only moderate dimensions"; and, second, "the moderate size of the average family impedes public awareness of growth and widely diffuses responsibility for it." In short, it is easily possible for population growth to keep right on rolling at a disastrous rate in advanced countries even when the average married couple believe that they are having a reasonable and moderate number of children. For this reason, the problem of excessive birth rates may prove in the long run to be even more intractable in the advanced countries than in the backward ones. One writer is of the opinion that if this problem is to be solved, governments will ultimately have to impose legal limits on the size of families, just as they now do on the speed of automobiles.

Well, whatever specific methods may be applied to achieve it, we know in general, thus, what the long-range solution to the population problem in both backward and developed countries must entail. This is the first of the two things that we can already say with certainty about man's chances of stabilizing his own numbers.

Decades of Crisis

The other is that there is almost no chance of achieving this goal within the next couple of generations. The world's population in 1960 was 3 billion, and by the end of the century — within your own lifetime — it will be more than six. This means that "during the second half of this century, there will be a greater increase in world population than was achieved in all the millennia of human existence up to the present time," in the words of Professor Philip Hauser of the University of Chicago. In the rather extensive reading that I have done on this subject, I have yet to find a single qualified authority who would disagree with this prediction or who thinks that there is anything practicable that we can do to prevent its fulfillment. True world population stabilization, if it comes at all, will not be achieved until sometime in the 21st century.

The general agreement on this suggests a last point with which to send you off to Saga. You and I will be retired before any real solution to the population problem can be achieved. Meanwhile, during the rest of our active lives, the world is bound to become an increasingly turbulent place in which to live. The doubling of man's numbers in the next forty years is going to subject all of his institutions and value systems to unbelievable strains. Specifically, we in the Western world are going to find it increasingly difficult to maintain our way of life in the face of outside pressures. The have-not nations today already comprise 66% of the world's population; by 2000 A.D. the figure will be 80%. In concrete terms this means that whereas today we already have two billion human beings who are receiving less than the minimum number of calories for mere subsistence, by 2000 A.D. there will be close to five billion starving people, *even if* we manage to double world food production in the meantime. One wonders whether we will still be able "four decades from now, to barricade ourselves in an oasis of relative plenty." Especially if by that time, as seems very likely, most of the non-Western nations have gone down the totalitarian drain and some of them have even acquired intercontinental missiles, as Red China will certainly have done. In the year 2000 will these nations still allow the United States, with less than 10% of the world's population, to get away with hoarding 43% of the wealth and consuming 50% of the non-renewable resources, as we did in 1950? I doubt it.

But this is only half the story. As the outside pressures on our way of life begin to mount, so will the pressure from within. The radical Right and the radical Left will both have their tawdry panaceas to peddle, ranging from preventive war to proletarian revolution. And I fear that more and more people, as reality becomes too much for them, are going to be seduced by the one or the other. They are going to experience what Gilbert Murray calls a "failure of nerve."

For a good many of you this is the last assembly of your college career. So I wish that I could send you out into the golden pre-commencement glow with a cheery, optimistic conclusion to this talk. But it would be dishonest of me to do that. I think that it is going to take lots of plain, old-fashioned guts for us to get through the rest of this century. It is going to be a world where, seemingly, a man can't step into the same river even once, and where there are no certitudes. Under those circumstances, I can only urge you to cultivate the most useful social grace in such a world: the ability to live with permanent crisis without panic.

Living Poor: 1968

Throughout this period, the little co-op was clearing jungle with machetes; I had promised that **AID** would help if they would work hard. By December, if everyone made a supreme and killing effort, we could move out of this stone-age agriculture, this farming with a machete and a sharp stick, into—well, into 1930 agriculture, at least.

But we had to clear enough land to make it worth while to bring the tractor out from Esmeraldas, across two rivers and up the thirty miles of beach. It would cost about a hundred dollars just to get the tractor to **Río Verde**, ready to begin

261

work. I was pushing the *socios* shamelessly, sulking when they promised to work and then didn't show up, getting mad at them when they quit early, pleading the case, giving them long, boring talks on the necessity of suffering Now, sacrificing Now. I was like a one-man symphony orchestra; when the flutes and violins didn't work, I was blaring trumpets and percussion. I tried to shame them or inspire them or obligate them in some way to cut down fifty acres of jungle. As I think back on that time, it occurs to me that they must all have thought I was quite mad.

There was hardly anything to eat in the town, and we were caught up in a monumental lethargy. The Italian priests who dominated the religious life of the province of Esmeraldas sent a fresh, plump brother out to take charge of the mission in Palestina. He had little money and to a large degree depended on the goodness of his parish; he lasted about four months, and when he was recalled he had lost about forty pounds and a good deal of his vocational calling. It wasn't that the people didn't want to feed him; it was simply that there was nothing to share, and many of the people were filled with shame and humiliation when the brother, vacant-eyed and ribs jutting, left the town.

It was the bananas that saved my life. When it was possible to buy them, I could generally manage to move around, but it meant eating bananas all day. (And I thought Ricardo was gassy!) Trying to set an example, I was clearing land on a daily schedule, and it became a fascinating problem in internal combustion to stuff bananas into myself and see how far I could go. Two bananas would get me up the hill to the farm; five bananas would fuel me up for forty or fifty minutes of low-keyed work; one banana would get me down the hill again to the Pepsi Colas and the animal crackers. When I went to work mornings, I had bananas stuffed into every pocket, pants and

shirt, the precise number counted out beforehand. Sixteen bananas would carry me through to noon if I didn't work too fast
or if the hordes of sons and grandsons of Sebastián Bagui
didn't shatter all my plans with their hungry cries as I passed
the last house on the street.

But being hungry wasn't simply losing my energy and
reaching a moment about eleven o'clock in the morning when I
ran out of energy and had to sit down every five minutes to
plan the next move, like a mountain climber at eighteen thousand feet. There was also a growing mental depression, a gray
fog of hopelessness that grew in my head each day; I could feel
myself getting stupider. Things became incomprehensible and
irritating. I snapped at the *socios* when they did dumb things,
even though I knew they were as hungry as I was and that they
were frantically combing the beach for something to give their
children.

I began to get furious letters from my friends and family
asking what was wrong, why didn't I keep in touch? I would sit
down, write "Dear Father," stare at the paper for fifteen
minutes, and then say, "Ah, to hell with it," and go take a nap.
It wasn't only that I couldn't think of anything to say; writing
a letter also involved finding the envelope, steaming it open,
addressing it, stamping it, getting it into the mailbox. The
whole thing was impossibly complicated.

Afternoons I usually stayed in the house and either slept or
just sat on a stool staring out the window at the ocean. I lay on
the bed and between naps read a book called *The Economics of
Subsistence Agriculture* in five-minute periods of comparative
concentration. Slowly I came to realize that the author was
writing about us in Río Verde when he said that a majority of
all the farmers in the world—perhaps 90 per cent of the self-
employed farmers in Africa, Asia, and South America—their
caloric intake limited to a bare subsistence level, worked no

more than three or four hours a day. There is only so much
energy in a dish of rice and a piece of fish. There are just so
many miles to a gallon of bananas—not one foot more.

I don't know why reading this in a book gave me such
satisfaction, seeing it all spelled out in graphs and statistics,
but it came to me as a revelation, this terrible truth that I had
known since arriving in Río Verde. And seeing it written down
wiped away my last lingering feelings that Ramón and his
two-hour naps after lunch, or Wilfrido sitting on a box all day
in the shadow of his house without moving, or all those
hundreds of men whose faces I had seen looking out the windows of little bamboo farmhouses up and down the river
through those long afternoons, were manifestations of laziness.
No, they were manifestations of exhaustion, in the case of
Wilfrido—an old man fading and aging before our eyes—a
moral exhaustion so consuming that he had to fight the impulse
to kill himself. I was making a hundred dollars a month. My
hunger was in varying degrees experimental and masochistic,
and resulted from laziness, bad planning, and affectation. I was
like a six-year-old kid playing doctor; I just wanted to see
what it was like.

But projecting my own lethargy, exhaustion, and mental
depression onto my friends, who weren't playing and who went
through this seasonal hunger every year of their lives, I began
to see in them such qualities of heroism and endurance, such a
wild and savage strength, that it about broke my heart with
pride for them.

Poverty isn't just hunger; it is many interlocking things—
ignorance and exhaustion, underproduction, disease, and fear.
It is glutted export markets, sharp, unscrupulous middlemen, a
lack of knowledge about the fundamental aspects of agriculture. It is the witchcraft of your grandfather spreading its
values on your life. It is a dozen irrational Latin qualities, like

your fear of making more of your life than your neighbor and thereby gaining his contempt for being overly ambitious.

There is no single way to smash out and be freed. A man has to break out in a dozen places at once. Most important, perhaps, he should start breaking out before he is six years old, for by then a typical child of poverty in a tropical nation is probably crippled by protein starvation, his brain dulled and his insides eaten up by worms and amoebas. No, more brutally true: if he is a typical child, an average child, by six he is dead.

To work harder a man has to eat better; to eat better he has to produce more; to produce more he has to work harder. And all of this is predicated on a growing knowledge of nutrition, basic hygiene, and the causes of the diseases that ravage his body; an understanding of agriculture and a respect for new farming techniques, new seeds, new ways to plant, new fertilizers, new crops.

Craziest and most interesting is the problem of incentive. Many of the people in Río Verde, for instance, aside from wanting more food, prettier clothes, and the money for doctors when they needed it, couldn't think of any good reason for not being poor. They didn't want anything. Perhaps a radio, perhaps a horse. To talk to a man about tripling his income to three hundred dollars a year was to fill him with confusion; he got nervous; he started to laugh; he wanted to go get drunk. The poor man from the moment of his birth was so inundated with problems, so deprived, that to end up wanting things was a form of insanity. What he wanted was to stay alive another day to tell jokes and visit with his friends in the sweet night air; he wanted new pants for the fifth of August fiesta, another pair at Christmas, and a house full of food for the Easter *Semana Santa;* he wanted ten sucres from time to time so that he could drink and dance and feel cleansed of life. Ramón with his composition roof was *egoísto,* the maverick; roofing a house with Eternit that would collect rain water, in this town of thatched roofs, had separated Ramón from the people. Ramón wanted a million things—a refrigerator, a larger house, a store-bought bed for the son he expected, and, not least, the respect of the middle-class storekeepers in Esmeraldas with whom he had done business all his life as just an undifferentiated shadow in the doorway, another beach *zambo*.

Ramón didn't want to be poor any more, and he was riding for a fall. The people had a growing contempt for his ambition and his aggressiveness, and he, a growing contempt for their lack of drive, their acceptance of the old ways. The time will come when he will have to find a middle-class environment where he can be at ease.

We cleared land until late November. By then I was aware that AID would probably not help us, or that if they did it would involve months and months of conferences with every gringo in the AID mission, from the office boy on up to the remote and invisible high mucky-muck. I needed the money in December, in thirty days. When the rains came our farm would turn into a mudhole, and the tractor would have to wait another year. It had been some time since I had pushed and raved at the *socios*. They had suffered, some of them, way past what anyone could reasonably expect, working without eating, sweating through the long morning hours. I was terrified that after all this work I wouldn't find the money, and I didn't know what I could say; they had too much faith in me. "Gringos never tell lies" was a sort of Ecuadorian truism, at least among the rural people who didn't know any better.

In December we burned the brush on the farm. We didn't have fifty acres cleared, but we had close to twenty, and it looked tremendous as it stretched away from the dry lagoons at the top of the hill down toward the ocean. For three weeks we burned brush, and the *socios* were exultant over what they had

done. Picking their barefoot way through the thorns and stick-
ers that covered the ground, they would pause in their work,
shade their eyes, play-acting, and look out over the ocean and
yell, "Here comes the *gabarrón* with the tractor; here comes
the tractor."

The shrimp boats were out there and the canoes of the
fishermen with their feed-sack sails, but there was no *gabarrón*
and there was no tractor, and after I talked for the last time
with **AID,** I knew that there never would be.

The Punishment of The Long-Distance Runner

The recent spate of track-record breaking, highlighted by 19-year-old
Jim Ryun's new world record for the mile, has provoked the usual amaze-
ment. How do athletes train themselves for such prodigious efforts? Is
there a limit to the speed at which men can run? How much of the effort is
physical and how much mental?

I have always taken the view that athletic barriers, like the 4-minute
mile, are largely psychological and I remember predicting the rash of sub-4-
minute miles after my own record in 1954. But I like to think that a 4-min-
ute mile is still an effort that makes our modern super-athletes slightly
breathless. At the time, I commented: "Apres moi, le deluge."

For eight years I had been training some five days a week for half an
hour, covering about five miles a day. This was my recreation from my
studies as a medical student in Oxford and in London. My training was based
on "interval running"--fast and slow running alternating. Each year, I gradually improved and in 1953 I had run a mile in 4 minutes 2 seconds, so that
I knew the 4-minute mile was within my compass. My chief anxiety was not
that I should fail to do it but that either Wes Santee in America or John Landy
in Australia would achieve the "dream mile" first.

Fortunately my chance came in May, 1954, in Oxford. Two other university runners, Christopher Brasher (later Olympic steeplechase gold
medalist) and Christopher Chataway (later Olympic finalist and defeater of the
Russian Vladimir Kuts), led in the early stages and ensured a sufficiently
fast pace for me to break the record. This rather cold-blooded attack on the
record gave me less satisfaction than later success against Landy in the
Commonwealth Games in Vancouver after he, too, had broken the 4-minute
mile.

The longer the distance an athlete runs, the more important is will power and the less important is physique. Let me illustrate this. In the 1952
Olympic Games, Emil Zatopek, the Czechoslovak "Iron Man" who set the
modern hard-training fashion, was running in his first marathon race, having
already won the 5,000-meter and 10,000-meter Olympic titles. The favorite
for the marathon was a very experienced English runner, Jim Peters.
Zatopek kept at Peter's shoulder for the first 10 miles. He then turned to
Peters and said in halting English, "Excuse me, Peters, but I have not run

268

a marathon before. Surely we ought to be running faster than this." Peters struggled on for a while and then collapsed completely.

The factors which limit speed of running differ with each athletic event and so the types of training needed to extend these limits also differ. The sprinter's speed depends on fast reaction time and innate speed of muscular movement, a quality that can be only slightly improved by training. This is the reason for the relative rarity of record-breaking in the sprint events. Charlie Paddock ran 100 meters in 10.2 seconds in 1921 and the world record is now only 0.2 seconds faster. Over the same period of time the mile record has fallen by some 15 seconds and the three-mile record by more than a minute.

The reason for the greater improvement in distance events lies in the effect of training on the body's capacity to transport oxygen from the air to the muscles. Sprinting avoids the need for external oxygen until the race is over by making use of the mechanisms described by doctors as "the oxygen debt," which cannot be increased by training--although the athlete with greater willpower can tolerate a larger debt.

The oxygen debt means the release of oxygen by the breaking down of energy-rich substances in the blood and muscles. The oxygen is provided at the cost of the production of a harmful substance, lactic acid. In large quantities, this substance causes the muscle pains and mental symptoms of severe fatigue. The lactic acid is gradually removed by oxidation after the exercise, during the phase of "breathlessness."

The greatest oxygen debts which athletes can tolerate are of the order of 20 liters of oxygen. This is more than enough to supply the energy for sprinting 100 yards but it is of only slight importance in a marathon.

However, the current uptake of oxygen by breathing air from the lungs and transporting it to the muscles can be "stretched" in a number of ways. The volume of the lungs increases with training and the membrane across which the oxygen diffuses probably becomes thinner. The heart becomes larger so that at each beat more blood--and hence more oxygen--is pumped to the muscles. Finally, the muscles themselves become better able to abstract oxygen from the blood.

In the mile, half the energy comes from the oxygen debt and the remainder from the transport of oxygen by the heart and lungs during the race. It is the current uptake of oxygen during exercise, with the more efficient use of the oxygen provided, which is therefore responsible for the improvement in middle and distance records over the last few years.

What is the ultimate in miling? Four minutes was only a milestone with the magical ring of a round number on the road to an absolute physiological limit. For athletes as they are now built this limit must be about three and a half minutes. Axiomatically, runners never reach the absolute limit, though they will approach it ever more closely. One could hazard similar theoretical limits, based on medical calculations of oxygen uptake and oxygen debt, for all distance races.

The last decade has seen the introduction of a new training technique to revolutionize previous concepts of the limit of oxygen uptake and use. Originated in Sweden before the war, it was called "Fartlek" or "speed play." It consists of running different distances from 100 yards to a mile at about three-quarters effort. A short period of time is allowed for partial recovery and then the effort is repeated, sometimes as many as 10 times.

The aim is to impose on the athlete the stress of oxygen lack repetitively and in this way increase the efficiency of the oxygen-transporting processes involving both heart and lungs. The better the athlete and the tougher his coach, the more of this training he can tolerate. Just to make things more difficult, Zatopek did his training in heavy army boots. Runners everywhere, without Zatopek's strength, gave themselves blisters and muscle strains trying to imitate him.

On average, Jim Ryun spent up to three hours daily in training of this kind. Sometimes he ran more gently before school in the morning, leaving the toughest session to the late afternoon. Obviously this kind of training requires a single-minded dedication, even ruthlessness, that few athletes possess. A few years ago, this grueling schedule was applied only in a mild form to schoolboy athletes because of the lingering fear of heart strain and other ills. We need not have worried, because the stress is repetitive, like the play of a child. The athlete can come to little physical harm, though he may, if the schedule is badly applied, become bored.

Recently, as the average age of record breakers has fallen, I have realized that another factor is at work. If "interval" training is started when the athlete is young enough (Ryun started when he was 16), his body can be "stretched" physiologically and anatomically to a degree that is impossible if the training starts after maturity when growth has ceased. This trend toward younger record breakers was first seen with 13-year-old swimmers whose lives under a Svengali-like coach consisted of little other than swimming. Perhaps only adolescents are sufficiently impressionable to believe that sport can be important enough to make this sacrifice worth while.

Will athletes eventually train for eight hours a day, virtually becoming running machines primed by food and drink, occasionally allowed to rest? For 5,000 meters, 10,000 meters and marathon runners, I predict that such schedules will eventually be devised, but I have still to be convinced that training of such severity is necessary for miling. Great though the margin of miling improvement has been in the last 12 years (nearly eight seconds) this improvement seems to me a rather poor return on a sixfold increase in the length of daily training. The optimum training for a miler may well be a judicious hour a day.

Even so, I am saddened by the thought that the place of the genuine amateur who earns a living in full-time employment elsewhere is bound to be less important in international athletics of the future. To put this another way, if I were a medical student today, I doubt if the cost of training, in terms of sacrifice of other pursuits, would attract me now to such serious athletics.

It is well known that the marathon represents the acme of athletic heroism. Here the athlete faces certain special hazards.

During the course of a six-mile race or marathon, especially with tem-
peratures in the mid-80's, the runner's muscles may generate more heat
than the athlete can dissipate. In spite of profuse sweating, the body temperature may rise to 105 degrees, the level of a malarial fever--just short of
the point at which heat stroke occurs. If the athlete has some trivial infec-
tion or if he fails to take enough salt and water, his sweating may fail and
within minutes he will collapse.

Many will have seen on television the most horrific marathon finish of
modern times. In the Commonwealth Games in Vancouver in 1954, Jim
Peters, the English marathon runner, tottered and sprawled over the last
lap. Officials were too bewildered to remove him from the track in his own
interests and in the interests of sport. Peters was hospitalized after the
race. He never competed again.

Few spectators ever realize the immense physical and mental strain of
such performances. An Australian spectator watching a grim marathon finish in the Melbourne Olympics was heard to exclaim, "Gee, the final of this
race should be worth watching."

The astonishing choice of Mexico City for the next Olympic Games in
1968 has introduced a new problem in distance training. At 7,500 feet there
is nearly 25 percent less oxygen in the atmosphere, and as we have seen,
performance is limited by the transport of oxygen. I do not agree with the
remark attributed to the Finish coach Onin Niskanen that "there will be those
that will die," but altitude could be the critical additional factor leading to
collapse under special circumstances.

There may be some athletes with heart abnormalities not previously detected, athletes with incipient infections or athletes who have simply not ac-
climated. Under conditions of Olympic competition, the mind sometimes
drives the body too hard and the normal protective mechanisms of fatigue fail
to operate. No instances of actual collapse have been reported from research
teams investigating the problem, but of course the true stresses of Olympic
competition cannot be reproduced. The conclusion I draw is that some risk
does exist and however small, it represents a powerful argument for never
holding distance events at high altitude.

However, one incidental result of the decision to hold the Olympic Games
in Mexico City has been the introduction of the idea of altitude training in order to improve the performance of athletes running either at high altitudes or
at sea level. Many of us had thought it a rather strange coincidence that the
only athlete who has won the marathon with ease in successive Olympic Games
is an Ethiopian, Bikile Abebe, who lives and trains at about 7,000 feet above
sea level. The point was reinforced recently when an unknown Kenyan runner,
Nestali Temu, who like Kipchoge Keino lives and trains at 6,000 feet, defeated
the Australian world-record holder Ron Clarke over 6 miles in the Commonwealth Games at sea level in Kingston, Jamaica.

After a moment's thought the implications are obvious. Performance is limited by the capacity to transport oxygen. Living and training at an altitude where there is less oxygen imposes a strain on oxygen-transport mechanisms that provokes an adaptive response and increases the athlete's efficiency. Already, several countries, with Mexico City particularly in mind, have established permanent training camps at high altitudes. I am worried that other countries, with more ruthless coaches, may well attempt to use low-pressure chambers in order to acclimate or train athletes artificially.

How far have we come from true sport? It would indeed be laughable if it were not so tragic. A pious rule from the International Olympic Committee limiting the length of altitude training to four weeks in the three months preceding the Olympic Games is a tacit admission of the blunder of holding the games at such an altitude and a recognition of the way in which athletes of the future will train.

It has been suggested, though I think without sufficient evidence, that there may be another new factor in record breaking--namely, the use of drugs. A few scientifically-controlled attempts to show that drugs consistently enhance an athlete's performance are unconvincing. For example, in one trial a group of athletes thought they recognized the effect of a supposed stimulant pill, but, in fact, the pill contained an inert substance. Some soccer players and cyclists, particularly on the Continent, have been thought to use drugs, but I think all track athletes realize that, besides being morally wrong, it would be the height of folly to imagine that they could consistently improve their performance by taking drugs.

There is as yet no proof that this is possible. The athlete must be able to compete consistently and within his own limits of recovery for many years.

The problem of training for field events has been helped by some technical advances. The invention of the fiberglass vaulting pole has modified styles and made all previous records obsolete. Analysis of styles of jumping, hurdling and throwing by means of motion pictures has enabled coaches to develop new techniques suited to the physiques of individual performers.

It is also commonly thought that a fine physique is an athlete's greatest asset. A British physician, Dr. J. M. Tanner, studied the physiques of more than 100 athletes at the Rome Olympics. Of course, a good physique is necessary, but Dr. Tanner concluded that the exceptional ability of these athletes lay less in their physiques than in the extraordinary physiological reserve produced by training and by their mental qualities. I can well recall seeing the 5-foot-6-inch English former world-record holder for the mile, Sidney Wooderson, effectively challenging the 6-foot-4-inch Jamaican Olympic champion, Arthur Wint.

All the changes in athletic training and performance in recent years seem to me to be part of a revolution in sport which has its explanation in individual and group psychology. Record breakers are athletes born with exceptional physical and mental qualities--freaks, in the complimentary sense. How else can we describe the capacity for mental excitement which brings with it the ability to overcome the discomfort, often pain, of extreme effort and to turn defeat into victory? One great athlete who trained himself mercilessly was Paavo Nurmi. With some insight into the psychological mechanisms under-lying such feats he made the bleak remark, "Only a poor man may run fast."

Only through considerable self-confidence and stoicism can the great athlete seem to bear lightly and with good humor the intolerable burden of being the favorite for a major race. It is this situation which Jim Ryun will now face. I can remember well the weeks before I raced against John Landy in Vancouver, when I felt that the eyes of the world were on us. My own expe-rience was suddenly stretched to the limits of what was tolerable. At this time, I discovered about myself both unsuspected weaknesses and also secret strength. A footballer's triumph or defeat is shared with other players, but the athlete must always stand alone. It was no comfort for me to reflect that men have faced greater trials or that it would all be the same in 100 years. For me, as for other athletes, the race is all-important, immediate and ir-revocable.

Dr. Kurt Hahn, the German who resisted Hitler's influence and came to England to embody his sociological beliefs in the founding of Gordonstoun, the school "adopted" by our royal family, has used an ambiguous phrase about sport's being the "moral equivalent of war." I think this is carrying the idea too far, but one cannot deny that man enjoys struggling to get the best out of himself. At one time the need for adventure may have been satisfied by the struggle for survival. Now, having conquered so many of the natural hazards facing him, man seeks further trials.

Thoreau once wrote, "The majority of men lead lives of quiet despera-tion." I believe that athletics, not necessarily at the international level, pro-vides an outlet for the craving for freedom which has become more important the more restricted, artificial and mechanized our society and work have be-come.

I think these psychological facts have provided the motive force behind the worldwide athletic revolution during the past decade. They have led to an enthusiasm for training for athletes without which no official encouragement or state interference would have had much effect.

Great athletic potential, like other unusual hereditary qualities, good or bad, occurs only in a small percentage of the population, perhaps less than 1 percent. Today, because athletic competition is universal in more and more schools, these oddities are uncovered by a new variant of the law of natural selection. The horrible thought of the eugenic selection of athletes has, I am glad to say, not so far achieved reality.

Thirty years ago, few had any chance to realize their potential; now, almost all those potential champions from an effective world population of some 800 million have their chance (and China has barely started to tap her immense physical potential). Moreover, besides the athletes recognizing their potential, a coach is likely to do so, too.

In underdeveloped countries, malnutrition and the struggle for basic essentials are still making the encouragement of athletic talent irrelevant, but this problem may be reduced in the next decade.

So sport, at one time the privilege of the few and the wealthy, has become the prerogative of the masses. That this should be so, and that for many the love of training, of varying severity, is better than mere watching, is necessary if our modern mechanized societies are to remain sane. Record breaking and yet stiffer training schedules will continue to be an expression of this social trend.

CALORIE EXPENDITURE THROUGH EXERCISE

Figured in calories expended per minute. For general activities, **figures for women slightly lower.**

Sleeping (basal metabolism) 1.0

Walking (2.5 mi. per hour) 3.6

Walking downstairs 5.2

Sitting in a chair reading 1.2

Sitting and eating 1.4

Standing up 2.3

Washing your face 2.5

Driving a car 2.8

Sitting and knitting 1.5

Sweeping 1.7

Preparing a meal 3.3

Scrubbing floors 3.6

Washing windows 3.7

Making beds 3.9

Ironing (standing up) 4.2

Washing clothes by hand 4.4

Hanging clothes on line 4.5

275

Painting at an
easel 2.0

Playing
piano 2.5

Bowling 4.4

Swimming
5.0

Golf 5.0

Ballroom
dancing 5.5

Horseback
riding (trot)
8.0

Skiing 9.9

Handball 10.2

Desk work
2.2

Shoemaker
3.2

Bricklayer
4.0

Carpenter
6.8

Farm work
in field 7.3

Mowing lawn
(hand mower)
7.7

Lumberjack,
chopping
wood 8.0

Salesman,
climbing
stairs with
sample case 9.0

THE GAS BUGGY IS HERE TO STAY

by JAY A. BOLT

The internal combustion engine
helped to shape
our industrial society. Now
the smogless automobile
is on the way, and nothing is
likely to take its place

On Thanksgiving Day in 1895, the first automobile race in America took place through the snow-covered streets of Chicago. Of the six vehicles that started, four had gasoline engines and two were battery-powered electrics. The winner was a "gasoline wagon" built by the Duryea brothers in Springfield, Massachusetts. It broke down twice but finally covered the fifty-four-mile course in ten hours and twenty-three minutes. Both of the electrics were forced to give up when their batteries ran down.

I am reminded of this bit of automotive history when I read predictions today that the electric automobile is about to return in vast numbers to the streets and highways of this country. I am an antique car enthusiast, and I admire the old battery-powered drawing room on wheels that was particularly popular with ladies in the years before and after the first World War. It was quiet and needed no cranking, whereas the balky gasoline buggies of the period required a strong right arm to get them started and mechanical wizardry to keep them running. They scared horses and woke people from their Sunday naps. But they won that long-forgotten race in Chicago, and they went on to sweep competing forms of transportation from the roads and turn rural America into a tightly knit, industrial society.

Renewed interest in electric-powered vehicles arises, of course, from widespread concern over air pollution. In the early nineteen fifties Dr. A. J. Haagen-Smit of the California Institute of Technology showed that certain ingredients of automotive exhaust react chemically in the presence of sunlight to form smog. Since then, a rapidly growing research effort has been focused on this problem. Legislation has been enacted in California and at the federal level requiring exhaust-control equipment on all new cars. Congress has held hearings on the subject of automotive air pollution, and conferences have been called by government and industry groups. I am serving on a technical committee that has been formed to advise the United States Department of Commerce and the Congress on the possibilities for reducing exhaust emissions from present automotive engines, and on possible alternative forms of vehicular propulsion. Many people have asked me what sort of vehicles I expect to see in our cities and on our expressways ten or twenty years from now.

Before making a prediction, let's take a brief look backward. At the turn of the century, when industrialization was bringing pressure for more flexible means of transportation than rails could provide, electricity, steam, and gasoline competed on a nearly equal basis for right of way on the open road. In some ways the electric vehicle was the most advanced; it had a dependable motor that operated smoothly and was easy to control. Steam cars had plenty of speed, power, and acceleration, but it took time to get them started. The gas buggy was something else again. Newspaper editorials denounced it as a menace to life and limb, and laws were passed restricting its use. But it had two overriding advantages: the fuel it used was inexpensive and could be supplied by any country store, and every teen-age boy who saw a gasoline automobile chugging down the road immediately wanted to own one. Once it was improved and available at a popular price, *everybody* wanted one. Enthusiasm for the gasoline automobile as the ideal form of private transportation has never lessened.

I have spent my professional life working with automotive power plants of all kinds, and it is noteworthy that the vast changes in the automobile, and the transformation it has brought to this country, have taken place within the span of a lifetime. In 1924, when I was thirteen years old, I acquired my first car without cost by pulling it home through the snow with a team of horses. It was obsolete and worthless—a one-cylinder, seven-horsepower Brush runabout built in 1908. It had two bucket seats, a remarkably modern coil-spring suspension, and a double-chain drive to the rear wheels, which were supported on an axle of solid hickory. A popular jingle said it had a "wooden frame, wooden axles, wooden wheels, and wouldn't run."

I made it run, though, and even poured hot metal to cast a new engine connecting rod bearing in the woodshed. The car would go about twenty miles per hour without shaking too badly, and it was a lot more fun than riding a bicycle.

It was cars like my old Brush and its successors that proved the superiority of the gasoline engine over electricity and steam. There are a few who steadfastly believe that if equal effort had been spent on steam or electric cars these would now be superior to the gasoline-engine-powered car. I do not agree. In comparison with electric cars, there was the simple matter of horsepower per pound. A Detroit Electric, one of the most popular town cars before World War I, weighed 5,500 pounds. A 1912 Model-T Ford touring car weighed 1,500 pounds.

Energy per pound of weight is still the major technical obstacle to the use of the battery-electric system for automotive propulsion because it severely limits the distance a battery-powered vehicle can go before requiring a recharge. One hundred pounds of gasoline—a tankful for most cars—will carry us about the same distance as a single charge supplied by 5,000 pounds of lead-acid storage batteries of the kind available today. Each recharge of the battery takes several hours compared to the few minutes required to fill a tank with fuel. It is obvious that the distance, speed, and performance characteristics of an electric are vastly inferior to those of a gasoline-driven car.

Despite their handicaps, battery-powered propulsion systems have been improved considerably over the years and have a number of practical uses. Battery manufacturers have had every incentive to upgrade their products, and in the past forty years the energy capacity per pound of the lead-acid battery has roughly doubled. A few electric delivery trucks are in use today in New York City, and thousands of electric milk and bread vans operate in Great Britain. They cover their daily routes of thirty miles or so at a speed of fifteen to twenty miles per hour and then go back to the garage for recharging overnight. Small cargo-hauling electric tractors and forklift trucks perform essential service in factories everywhere. And of course the most popular electric vehicle for passengers is the golf cart, which can roll twice around the course on a single battery charge.

It is from this base—the current "state of the art"—that we must consider the future of electric automotive power. From even a much-improved golf cart it is a long step to a practical road vehicle. Moreover, considering the time required to bring a proven prototype model of a new and very different automobile to mass production, the electric passenger car of 1975 should be running on a test track today. A vehicle that might be considered a substitute for the present gasoline automobile is nowhere in sight, in spite of substantial expenditures of money and effort by industry in the United States and abroad, and in spite of newspaper articles I have seen with photographs of "the new look in electric cars."

Before a practical electric automobile of conventional size can be produced and sold, a major technical breakthrough in battery research must take place. One of the best experimental electrics has been demonstrated by General Motors engineers, who equipped one of their standard production cars with a very high performance experimental 115-horsepower electric motor and a silver-zinc battery pack weighing 660 pounds. Silver-zinc batteries have about five times more energy storage per pound of weight than lead-acid batteries, and with the advanced type of motor and controls developed for the car, its performance was made comparable to a gasoline-powered automobile. But the annual world production of

silver would only be enough to manufacture 120,000 cars of this kind—and the batteries alone would cost thousands of dollars each. General Motors has stated that this is obviously not a commercial solution of the problem.

The Ford Motor Company is at work on a new battery that uses sodium and sulfur as the basic elements. Although it is said to have about fifteen times the energy storage capacity of a standard lead-acid battery, it is in a very early stage of development. In order to operate, it must be kept at a constant temperature of at least 500 degrees Fahrenheit, a formidable safety hazard in practical use. At General Motors similar research work is in progress on high-temperature, high-energy batteries. The Edison Electric Institute and General Dynamics Corporation are jointly spending several million dollars on the development of a high-energy zinc-air battery system, but its proponents cannot predict when it will be ready to power a car comparable to today's standard family sedan. These are worthwhile research activities and should be viewed with optimism, but their practical usefulness is still uncertain.

Perhaps the most intriguing type of vehicle power system in terms of future potential is the fuel cell. It also presents the greatest problems in practical application. Unlike a storage battery, a fuel cell converts chemical energy to electric energy continuously so long as it is refueled and thus does not need to be recharged. Like most other power plants, it is not a recent conception. Sir William Grove, an Englishman, demonstrated about 1840 that hydrogen and oxygen could be combined with the aid of catalytic platinum electrodes to produce water and a small electric current. This is essentially the process that was developed and improved at great cost to generate electricity for the Gemini and Apollo space vehicles. The fuel cell system in the Gemini spacecraft weighed about 400 pounds and delivered one kilowatt, or one and one-third horsepower.

Before there can be any real commercial application of fuel cells for automobile propulsion, their cost must be greatly reduced and their power output greatly increased; the platinum in the Gemini unit was worth approximately $7,500. For an automotive power plant, the liquid hydrogen and oxygen used in a spacecraft present serious problems, but petroleum-base fuel that is cheap and safe to handle is now being burned in small experimental cells. One such fuel cell generator, with enough power output to drive a small power tool such as an electric drill, is in the developmental stage at the laboratories of the Esso Research and Engineering Company. General Motors has built an experimental Electrovan run by fuel cells that consume liquid hydrogen and oxygen, but the propulsion equipment weighs 3,650 pounds and takes up most of the space inside the vehicle.

Fuel cells, despite their present shortcomings, may be a better long-range hope than storage batteries as a means of vehicular propulsion. When will we have fuel cell automobiles? An answer was recently given to a Congressional committee by Dr. Arthur M. Bueche of the General Electric Company. "I don't know *when,* and I'm not really sure *if.* All I can say is: on the basis of what we know now, fuel cells *might* someday be very attractive for vehicles because they *might* not give off any appreciable noxious exhaust, and they *might* be developed to fit into more compact portable packages, and they *might* be made inexpensively enough for general use in vehicles."

Personally, I don't believe we will have to wait for all these "mights" to be resolved before we have motorcars that will reduce automotive air pollution to acceptable levels. Let's look at what has happened to the familiar gasoline-burning car in the last few years. In California, where the geography and weather conditions in Los Angeles County create unusually severe air pollution problems, the first devices to reduce automotive emissions were installed in 1961 model cars,

and two years later on all new cars sold in the United States. These devices controlled one of the three sources of hydrocarbon emissions from an automobile: the crankcase blowby. A second source is unburned fuel in the exhaust, and these emissions are being reduced by modification of the engine itself and by systems that burn most of the escaping hydrocarbons in the exhaust manifold. The third source, evaporative loss from the fuel system, will probably be bottled up soon by techniques such as the one announced recently by the Esso Research and Engineering Company, which uses activated charcoal to soak up hydrocarbons from the fuel tank and carburetor bowl.

In addition to unburned hydrocarbons, other pollutants formed in the exhaust of gasoline engines are carbon monoxide and nitrogen oxides. Intensive study has been undertaken in recent years to discover just how these undesirable emissions are formed and how they can be eliminated. I learned long ago that much that happens in an automobile engine has never been completely analyzed or understood. Early gasoline engines were designed mainly by men who were skillful mechanics and machinists rather than engineers with a strong scientific background. They didn't have to know much about theories of thermodynamics and combustion to build a car that would run well, and there was little incentive to worry about what came out of the tail pipe.

Today, with the urgent need to reduce exhaust emissions, the most sophisticated scientific instruments and techniques are being used to analyze the chemical processes that take place in an engine, so that these processes can be understood and controlled. By more precise regulation and distribution of the air-fuel mixture, for example, it is possible to burn the fuel more completely and to reduce greatly the hydrocarbons and carbon monoxide from the exhaust. In present practice, richer than ideal air-fuel mixtures are used under many operating conditions. But important improvements can be made in this area. Less pollution and greater fuel economy will result.

The automotive laboratories in Detroit are using instruments today that can measure exhaust components in parts per million. The mass spectrometer is being applied to the difficult task of detecting and measuring oxides of nitrogen. Other instruments that did not exist ten years ago are now in routine use. Engine intake manifolds, which for more than half a century have been designed largely on a trial-and-error basis, are being more carefully tested, analyzed, and re-engineered. The fundamental characteristics of carburetors, and the complex flow phenomena that occur in engine induction systems, are subjects of current cooperative research by industry, the universities, and the United States Public Health Service. The new knowledge being acquired will permit the design of improved carburetors and induction systems that will more nearly supply the optimum air-fuel mixture to all engine cylinders at all times.

Combustion chamber deposits have been found to affect the amount of fuel left unburned, so that the use of high-quality gasolines and oils, with additives that minimize the formation of deposits, has taken on new importance.

While all practical steps are being taken to control emissions by means of engine design improvements, exhaust impurities are being further reduced through the installation of special accessories in the exhaust system. Such devices are already in use in California, and potentially superior types of systems are under development. These include improved versions of the air injection exhaust reactor now being used in California, and other systems such as the direct flame afterburner and the catalytic converter. Both have as their objective the oxidation of carbon monoxide and unburned hydrocarbons. A method of reducing up to 85 per cent of the oxides of nitrogen—the third group of undesirable exhaust ingredients—was shown recently to a Senate subcommittee that is considering new air pollution legislation.

These developments make it reasonable to predict that by 1975 automobiles will be manufactured that emit only 10 per cent of the hydrocarbons and carbon monoxide produced by an uncontrolled car of recent vintage. Reductions of this magnitude should end serious pollution of the air by automobiles over most of the United States for the foreseeable future, allowing for the anticipated growth in the number of vehicles. I am confident that within twenty years further engineering advances will lead to essentially pollution-free internal combustion engines.

Is there no future, then, for the vehicles propelled by the new batteries and fuel cells that are being designed, tested, and discussed so widely today? I think there is, when we consider the changing patterns of urban development and the population growth and transportation needs of tomorrow. A small electric passenger car, which we can visualize as a kind of glorified golf cart, could very well fill a need for short-range, low-speed travel from home to supermarket, school, or railroad station. On crowded city streets or local thoroughfares set aside for their use, small electrics could maneuver easily and silently. Delivery trucks that cover regular routes and return to central recharging stations might be economically attractive with the improved batteries that are on the way.

We can look forward to many strange new kinds of motorized transport on the streets and highways of the world by the end of this century. One forecaster has suggested that the average family will need a five-car garage to store all its vehicles. There may be battery tricycles and runabouts sharing space with the new smogless, fume-free sports cars, passenger cars, and station wagons. There will certainly be turbine-powered trucks, buses, and bulldozers doing heavy work of all kinds. The "gasoline wagon" will, in my opinion, have no serious competition for ordinary road travel. It will still be king of the highway, just as it was in that historic motor race in Chicago seventy-two years ago, when a horse could have beaten the winner.■

Chapter Seven

NUCLEAR ENERGY: THREAT OR PROMISE

The atomic nucleus has become a symbol of the danger facing mankind from an uncontrolled arms race. We live in an age of massive deterrence, with each superpower believing that the other is prevented from unleashing an attack only by the threat of reprisal in kind. The missiles that can deliver nuclear warheads are becoming ever more powerful, more numerous, and more highly automated. Yet peace seems to be more brittle than ever.

How did it all begin? Very early in this century Albert Einstein proposed a new view of space and time, his Special Theory of Relativity. One of the consequences of this view was the prediction that all matter, as a result merely of existing, was a form of stored energy. The stored energy was enormous by usual standards--one kilogram of matter could supply all the energy requirements of a city of half a million inhabitants for a whole year. For Einstein at this time, however, the result was of scientific interest alone, and he did not foresee the actual conversion of matter into energy on a significant scale. You must remember that the nuclear model for the atom was invented several years later!

What made the crucial difference arose from certain mysterious chemical observations on radioactive materials including uranium, observations that led to the concept of nuclear fission (so named in analogy to bacterial fission). As described by O. R. Frisch, in the first selection of this chapter, the liquid drop model for the nucleus provided a point of view from which nuclear fission was a plausible process. Even though only a very small fraction of the nuclear matter is converted to energy, the amounts were much larger than any previously conceived. This development was sufficient to alert the scientific community to the military implications of nuclear energy, and Albert Einstein was persuaded to write to President Roosevelt, calling his attention to the possibilities. Their fateful exchange of letters is the second selection in this chapter.

After considerable delay, the Manhattan Engineer District under Major General Leslie R. Groves was launched and feverishly pursued its mission until the Los Alamos Laboratory, led by J. Robert Oppenheimer, was ready to test the first bomb in July, 1945. At the same time, planning for the potential use of the bomb, if successful, also went forward in the governmental councils. The test at Alamogordo is described vividly by Brigadier General

Thomas F. Farrell, who observed it in the company of the laboratory's scientific leadership from a shelter about six miles from the point of the explosion.

Following the successful test, the Allies issued the Potsdam declaration, demanding that Japan surrender or face destruction. When Japan rejected this ultimatum, President Truman authorized dropping atomic bombs on Hiroshima and Nagasaki in early August. He was motivated by the prospective cost in human lives of an Allied invasion of the Japanese Islands. Japan offered to surrender and accepted the Potsdam terms a few days later.

The bomb had been developed and used under wartime pressure, with the objective of ending the war overriding all other considerations. After the surrender of Japan, mankind faced the problem of living with nuclear energy and nuclear bombs as realities rather than as theoretical speculations. The United States Government responded to this situation by establishing a civilian agency, the Atomic Energy Commission (AEC), to control all facets of nuclear energy development for weapons and for peaceful applications. The scientific community responded by establishing the Federation of American Scientists, a new professional society, and publishing the Bulletin of Atomic Scientists, a journal devoted to publicizing and debating problems of the nuclear age.

The pages of the Bulletin speak eloquently of man's efforts to achieve international control of nuclear energy, the continued weapons development during the cold war, loyalty and security investigations, and the achievement of a nuclear test ban treaty. The Atomic Age, edited by Morton Grodzins and Eugene Rabinowitch (New York: Simon and Schuster, 1965) contains a broad selection of articles from the Bulletin and dealing with these topics.

The remaining contributions in this chapter present a very brief survey of developments since the Alamogordo explosion. J. Robert Oppenheimer presents his somber views shortly after the war's end, and Andrei D. Sakharov reiterates these more than twenty years later.

Is there no progress, then? Nuclear energy for peaceful purposes has been developed extensively and promises to contribute significantly to man's energy needs in the future. Yet the hazards associated with the intensively radioactive byproducts continue to generate controversy. Several avenues of application are being pursued. One of these is the use of the radioactive materials themselves for food preservation, medical treatment, and chemical research. A second is the use of nuclear reactors to operate electric power generating plants. Edward Teller (for many years on the AEC's Committee on Reactor Safeguards) and Albert Latter review the safety problems associated with such plants. A third avenue is the channeling of the energy from nuclear explosions so as to accomplish large scale earth movement or excavation. This is being studied in the AEC's Plowshare Program, which is described by Gerald W. Johnson.

About the Contributors

OTTO ROBERT FRISCH (1904-   ) is Jacksonian Professor at the University
of Cambridge. His experimental work has contributed substantially to current models for the atomic nucleus. In addition to research papers, Professor Frisch has written several books explaining physics to the layman.

ALBERT EINSTEIN (1879-1955) was one of the world's greatest scientists
and has come to symbolize modern physics in the public mind. He taught at
several European universities before he joined the Institute for Advanced
Study, Princeton, N. J. in 1933. Einstein had a deep compassion for op-
pressed people and became an ardent supporter of world government after
World War II.

FRANKLIN DELANO ROOSEVELT (1882-1945) was thirty-second president
of the United States. He was elected to four terms and served from the
Great Depression until close to the end of World War II.

THOMAS F. FARRELL (1891-1967) was Brigadier General and Deputy to
Commanding General Leslie R. Groves of the Manhattan Engineer District.
General Farrell was trained as a civil engineer. He served in the army
during both world wars, earning the Distinguished Service Cross, the French
Croix de Guerre with Palm, and the Distinguished Service Medal for heroic
and meritorious service.

J. ROBERT OPPENHEIMER (1904-1967) was a distinguished theoretical
physicist, specializing in quantum theory, who directed the Los Alamos
Laboratory where the atomic bomb was designed. After the war he became
director of the Institute for Advanced Study, Princeton, and also served as
Chairman of the General Advisory Committee of the U. S. Atomic Energy
Commission. In 1963 he received the Commission's Enrico Fermi Award
for his outstanding contributions to the development of nuclear energy.

ANDREI D. SAKHAROV (1921-   ) is a theoretical physicist at the Lebedev
Institute of the USSR Academy of Science, Moscow. His work has included
contributions to the harnessing of thermonuclear energy.

EDWARD TELLER (1908-   ) is Professor-at-large at the University of
California. He is a theoretical physicist, specializing in molecular and
nuclear physics. At the Los Alamos Laboratory he made decisive contributions to the development of the hydrogen bomb. After the war he returned to
university teaching but continued to serve the Atomic Energy Commission
and the Defense Department. For his scientific work and public service he
was awarded the Commission's Enrico Fermi Award in 1962.

ALBERT L. LATTER (1920-   ) heads the Physics Department of the Rand
Corporation, Santa Monica, California. He is a theoretical physicist whose

research has been concerned with the properties of matter under the conditions of a nuclear explosion. Dr. Latter has worked on the design of nuclear weapons, participated in test ban negotiations, and sought to inform the public about potential benefits from nuclear energy development.

GERALD W. JOHNSON (1917-   ) is an experimental physicist and manager, Explosives Engineering Services, for Gulf General Atomic, Inc. Prior to taking this position, Dr. Johnson was associated for many years with weapons development and the search for peaceful uses of atomic energy by the Atomic Energy Commission.

Atomic Nucleus

by Otto Robert Frisch

The first time I met Niels Bohr was in 1933, in Hamburg. Hitler had come to power, and Bohr was travelling about and talking to his colleagues in order to find out how many German physicists would be dismissed under the new race laws, and how one could best organize help for them. I had just succeeded in measuring the recoil which a sodium atom suffers on sending out a light quantum. To me it was a great experience to be suddenly confronted with Niels Bohr—an almost legendary name for me—and to see him smile at me like a kindly father; he took me by my waistcoat button and said: "I hope you will come and work with us some time; we like people who can carry out 'thought experiments'!" You see, the recoil of an atom had been discussed a good deal in the early days, but few people had thought that it could be measured. By the time I did that it was no longer difficult; Otto Stern—under whom I worked in Hamburg —had created the techniques that were needed.

That night I wrote home to my mother (who was naturally worried about my future, in view of Hitler's race laws) and told her not to worry: the Good Lord himself had taken me by the waistcoat button and smiled at me. That was exactly how I felt.

During the next year I worked in London with Blackett where I learned the techniques of nuclear physics. A few months later the great discovery was made in France that new radioactive bodies could be made artificially by bombarding ordinary light elements with alpha particles; I quickly jumped on that bandwagon and was lucky enough to be one of the first to publish further results. I also played around with Wilson chambers and constructed one which was easy to build and convenient for visual observations. And when my stay in London drew to a close I found myself with an invitation to go and work in Copenhagen, in Niels Bohr's Institute.

Arriving in a new place is always a confusing business, and I have no clear recollection of it. There seemed to be a lot of odd people. There was Placzek, always sleepy (except late at night) and usually unshaven; it took me a long time to discover what a brilliant mind there was behind that bohemian façade (he was actually from Bohemia). He took me to hear a talk by Gamow on the very first day; the talk was supposedly in Danish, but I had no trouble understanding the peculiar Esperanto which Gamow spoke (whichever language he was supposed to be speaking). From one of the early colloquia the scene of a discussion between Bohr and Landau is imprinted on my mind: Bohr bending over Landau in earnest argument while Landau gesticulated at him, lying flat on his back on the lecture bench (neither seemed to be aware of the unconventional procedure). After my three years in the conventional atmosphere of Hamburg and one year in London, where I felt shy and made few contacts, it took me quite a time to get used to the free and easy atmosphere at Blegdamsvej 15 where a man was judged entirely by his ability to think clearly and honestly.

Bohr in those days seemed at the height of his powers, bodily and mentally. When he thundered up the steep staircase, two steps at a time, there were few of us younger ones that could keep pace with him. The peace of the library was often broken by a brisk game of ping-pong, and I don't remember ever beating Bohr at that game. His reactions were very fast and accurate, and he had tremendous will power and stamina. In a way those qualities characterized his scientific work as well.

My own work did not give me much contact with Bohr; I was concerned with the construction of simple and reliable Geiger counters and auxiliary equipment, and I continued my hunt for new radioactive elements (I found two more). But Bohr often wandered into the laboratory to watch us and clearly wanted to help. On one occasion he even offered help, assuring us that he was not as clumsy as he looked, and before we could stop him he had picked up one of the special thin-walled counters which were made for the study of rays of low penetrating power. The counter immediately crumpled up with a nasty crackling sound, and Bohr went out of the room looking very embarrassed. Yet how could he have known that such a counter could barely stand the pressure of the air and was sure to collapse even under the gentlest touch?

But of course we went to the colloquia which took place at frequent intervals and were often called together at very short notice, and there Bohr's unique powers came into full play. Many have found Bohr's writings involved and indirect and hard to follow; but in a discussion he was direct and vigorous. Of course he was far too kind to tell anyone that he was speaking nonsense, but the meaning of when Bohr said "very, very interesting" was soon clear to all of us. The idea of complementarity was the subject of many discussions. I remember one occasion when someone said at the end that it made him quite giddy to think about those questions; Bohr immediately replied: "But if anybody says he can think about quantum problems without getting giddy, that only shows that he has not understood the first thing about them."

From time to time there were crises. I remember the paper by the USA experimenter Shankland who claimed to have shown that the conservation of energy did not hold in collisions between high-energy photons (gamma quanta) and electrons. That was very disturbing; if true it would have thrown physics back to a much earlier stage when Bohr himself had thought (mistakenly, he now felt sure) that the energy law was true only on the average and not necessarily for the individual event. Shankland's result gave rise to agonizing discussions which made it ever clearer that it could not be reconciled with the advances that physics had made since those early days. It was a great relief when in the end it became known that Shankland had overlooked certain sources of error and that energy was in fact conserved in each single collision.

So it was always: every inconsistency was an enemy who had to be at once attacked, and against whom Bohr turned the full strength of his powerful mind. Sometimes, as we have seen, the experiment turned out to be wrong; sometimes Bohr would tell us one day, with even greater delight, that he had made a mistake, that the inconsistency disappeared when one found the right way to think about it. He was deeply disturbed when he heard that the impact of a fast proton on a lithium nucleus sometimes caused the emission of a hard gamma ray; he felt sure that the nucleus formed by such an impact would break up into two helium nuclei much too quickly to allow a gamma ray to be sent out. I still remember how delighted Bohr was when he realized that the nucleus could be formed in a state which—because of the symmetry laws of quantum theory—

could not break up into two helium nuclei. He never hesitated for a moment to admit that he had been in error; to him it merely meant that he now understood things better, and what could have made him happier?

Soon after my arrival came the exciting news that Fermi in Rome had found that a great many new radioactive nuclei could be made by bombarding ordinary nuclei with neutrons. I was one of the few who could read Italian, and I well remember how everybody crowded around me whenever a new issue of "La Ricerca Scientifica" had arrived, the small journal to which Fermi had granted the privilege of printing his discoveries because there they were printed quickly. It was at once clear that it was very important to acquire a source of neutrons and the chemist Hevesy (Bohr's friend and collaborator during many years) appealed to the Danish people for a sum of 100,000 Danish Kroner, to buy half a gram of radium for Bohr to his fiftieth birthday. To turn this into a neutron source it had to be mixed with several grams of beryllium powder, and it became my task to provide that. Beryllium is a light and silvery but very hard metal, and I got almost everybody in the Institute to help grind it up in mortars, which was many days of hard work. In those days it was not known that some people are ultrasensitive to beryllium and would probably have died from breathing traces of beryllium powder during the grinding. But fortunately none of us was, and no accidents happened.

So, from 1935, the Institute had a strong source of neutrons available. Most of the neutrons were used to make radio-phosphorus for Hevesy and his people, to be used in biological tracer experiments. For that purpose the source was kept in a large flask containing about ten litres of carbon disulphide, a toxic and highly inflammable liquid; for safety, the flask was kept at the bottom of a deep well right inside the Institute. Every two weeks the liquid was changed, and the irradiated liquid was processed to extract the radio-phosphorus, P-32. On one occasion the bottle slipped and broke, and the assissant escaped up the spiral staircase, pursued by the fumes which would have blown up a large part of the Institute had there been a spark. Our largest pump was set up to exhaust those fumes, but even so it took all night. In those days I lived at the top of the Institute, and I remember going to bed in a fatalistic mood. However, the next morning I was still there and the danger was over.

In the meantime, stranger and stranger news came from Rome. Fermi had found that neutrons could be slowed down by having them pass through hydrogen-containing materials such as paraffin wax or water, and that such slow neutrons were very easily captured by nuclei. That was very surprising. According to what was then believed about nuclei, even a slow neutron should usually pass clean through a nucleus, with only a small chance of being captured. Bethe in USA had tried to calculate that chance, and I remember the colloquium—late in 1935—when someone reported on Bethe's paper. Bohr kept interrupting, and I wondered, a bit impatiently, why he didn't let the speaker finish. Then, in the middle of a sentence, Bohr suddenly stopped and sat down, his face suddenly dead; we feared he had been taken unwell. But after only a few seconds he got up again and, with his apologetic smile, he said: "Now I understand it."

The understanding he reached in that memorable colloquium has become known as the "compound nucleus". Bohr realized then that a neutron on entering a nucleus at once collides with one of the protons or neutrons inside, and that those particles in turn suffer collisions so that the original energy of the neutron is very quickly distributed among many particles. After that it takes a long time (really a minute fraction of a second, but long by nuclear standards) before one of the particles happens to get enough energy to escape, and if during that time a photon (a gamma quantum) is sent out, the neutron can no longer escape. The long time during which the "compound nucleus" exists meant that—because of the uncertainty principle—it could possess sharply defined energy states and that was an exciting prospect.

I didn't understand much of that at the time, but Placzek did. Together we made measurements of the absorption of slow neutrons in gold, cadmium and boron, and various combinations of those elements. There were good reasons to think that the absorption of neutrons in boron should follow a simple law, namely that it should be inversely proportional to their speed, and from our measurement Placzek could work out the speed of those neutrons most strongly absorbed in gold. The result was that gold showed a sharp "resonance" for neutrons of only a few volt energy. That was many thousand times less than anyone had expected previously, but just what Bohr had come to expect from his idea of the compound nucleus. You can imagine how pleased Bohr was about the outcome of our work, and that he urged us to publish it quickly. That was not so easy. Placzek and I had very different ideas on how to present our results, and since he only woke up in the evening we had to do our writing at night and most of the final discussions in the small hours when I was sleepy and stubborn. But after a few strenuous nights the text was finally agreed on, despite Placzek's passionate protestations, and I took it to the post office myself at four in the morning, to prevent any resumption of the argument.

By the way, for those measurements we needed quite thick layers of gold, and it was Placzek's inspiration to use those several Nobel Prize medals which some of Bohr's friends had left with him for safe keeping when the Nazis came to power. It was a source of great satisfaction to us that those otherwise useless medals should thus be employed for a scientific purpose! The medals were once more in danger when Denmark was occupied by the Nazis and were then saved by Hevesy, who dissolved them in acid and kept them in a bottle. After the occupation, they were turned into gold again and struck as medals once more. But that was after my time.

The other person with whom I was in close contact at that time was Laslett, a slim, bony, taciturn American who had come to help us build a cyclotron. He seemed to spend much of his time in a chair tilted back, his feet on the table, a drawing board on his lap, unperturbably designing one component after another. The decision to build a cyclotron had been taken in 1935. I remember when the 40-ton magnet, a gift made to measure by Thomas B. Thrige in Odense, arrived we found that its largest piece was too big to go through the window into the room built for it, so a part of the wall had to be knocked out. Two other cyclotrons of similar size were being built at the same time in Stockholm and Liverpool; we had a good deal of correspondence with those two places and once paid a visit to Stockholm where we "Kärngubbarna" were received with true Swedish hospitality. In the end it was our machine that was the first to work, and thus the first cyclotron on this side of the Atlantic.

In addition it was decided to buy a high-voltage generator for one million volts from Germany, and I was sent to Leipzig to settle some of the details with the makers. It gave me a sardonic pleasure to be treated as an honoured guest by the people who had sacked me under Hitler's race laws but who were willing to forget those laws when it was to their own financial interest. The machine was duly delivered, and it did work when I left Copenhagen.

Those were happy days for me. The discovery of neutron resonances had opened a new field for research where interesting experiments could be done quickly and with simple means. Today, of course, dozens of resonances have been mapped out with great accuracy for every element on the chemists's shelf; but in those days it was very exciting when in a few selected elements—gold, iodine, arsenic—the first resonance could be roughly located and its main characteristics determined. All this served to support Bohr's concept of the compound nucleus and to teach us more about its properties, and Bohr naturally took greater interest in that work. But the most personal contacts we had on those frequent occasions when Bohr had invited a number of us out to Carlsberg where, sipping our coffee after dinner, we sat close to him—some of us literally at his feet, on the floor—so as not to miss a word. Here, I felt, was Socrates come to life again, tossing us challenges in his gentle way, lifting each argument onto a higher plane, drawing wisdom out of us which we didn't know was in us (and which, of course, wasn't). Our conversations ranged from religion to genetics, from politics to art; and when I cycled home through the streets of Copenhagen, fragrant with lilac or wet with rain, I felt intoxicated with the heady spirit of Platonic dialogue.

In 1938 the horizon began to darken. We had all been watching Nazi Germany with apprehension, feeling that her philosophy would lead her to a war of conquest before long. Her occupation of the Rhineland made us fear that the time would not be long now. The occupation of Austria in March 1938 changed me—technically— from an Austrian into a German. My aunt, the physicist, Lise Meitner, who had acquired fame by many years' work in Germany, now had to fear dismissal, and there was also a rumour that scientists might not be allowed to leave Germany; so she decided to leave secretly, assisted by her friends in Holland, and in the autumn she accepted an invitation to work in Stockholm, at the Nobel institute. I had always kept the habit of celebrating Christmas with her in Berlin; this time she was invited to spend Christmas with Swedish friends in the small town of Kungälv (near Gothenburg), and she invited me to join her there. That was the most momentous visit of my whole life.

Let me first explain that Lise Meitner had been working with the chemist Otto Hahn for about thirty years, and during the last three years they had been studying the radioactive bodies that were formed from uranium under bombardment with neutrons. Fermi, who first discovered those bodies, thought they were new 'transuranic elements', that is, elements beyond uranium (the heaviest element found in nature), and Hahn and Meitner at first got results that confirmed that view. But it seemed difficult to account for the large number of substances formed, and things got even more complicated when some were found (in Paris) that were apparently lighter than uranium. Just before Lise Meitner left Germany, Hahn had confirmed that this was so, and that three of those substances behaved chemically like radium. It was hard to see how radium—four places below uranium—could be formed by the impact of a neutron, and Hahn decided to carry out thorough tests in order to make quite sure that those substances were indeed of the same chemical nature as radium.

When I came out of my hotel room after my first night in Kungälv I found Lise Meitner studying a letter from Hahn and obviously very puzzled by it. I wanted to discuss with her a new experiment that I was planning, but she wouldn't listen; I had to read that letter. Its content indeed was so startling that I was at first inclined to be sceptical. Hahn and Strassmann had found that those three substances were not radium, chemically speaking; indeed they had found it impossible to separate them from the barium which they had added in order to facilitate the chemical operations. They had come to the conclusion, reluctantly and with hesitation, that they were isotopes of barium.

The suggestion that they might after all have made a mistake was waved aside by Lise Meitner; Hahn was too good a chemist for that, she assured me. But how could barium be formed from uranium? No larger fragments than protons or helium nuclei (alpha particles) had ever been chipped away from nuclei, and the thought that a large number of them should be chipped off at once could be dismissed; not enough energy was available to do that. Nor was it possible that the uranium nucleus could have been cleaved right across. Indeed a nucleus was not like a brittle solid that could be cleaved or broken; Bohr had stressed that a nucleus was much more like a liquid drop. Perhaps a drop could divide itself into two smaller drops in a more gradual manner, by first becoming elongated, then constricted, and finally being torn rather than broken in two? We knew that there were strong forces that would resist such a process, just as the surface tension of an ordinary liquid drop resists its division into two smaller ones. But nuclei differed from ordinary drops in one important way: they were electrically charged, and this was known to diminish the effect of the surface tension.

At that point we both sat down on a tree trunk (all that discussion had taken place while we walked through the wood in the snow, I with my skis on, Lise Meitner without), and started to calculate on scraps of paper. The charge of a uranium nucleus, we found, was indeed large enough to destroy the effect of the surface tension almost completely; so the uranium nucleus might indeed be a very wobbly, unstable drop, ready to divide itself at the slightest provocation (such as the impact of a neutron).

But there was another problem. When the two drops separated they would be driven apart by their mutual electric repulsion and would acquire a very large energy, about 200 million electron-volts in all; where could that energy come from? Fortunately Lise Meitner remembered how to compute the masses of nuclei from the so-called 'packing fraction formula', and in that way she worked out that the two nuclei formed by the division of a uranium nucleus would be lighter than the original uranium nucleus by about one fifth the mass of a proton. Now whenever mass disappears energy is created, according to Einstein's formula $E = mc^2$, and one fifth of a proton mass was just equivalent to 200 million electron-volts. So here was the source for that energy; it all fitted!

A couple of days later I travelled back to Copenhagen in considerable excitement. I was keen to submit our speculations—it wasn't really more at that time—to Bohr, who was just about to leave for the USA. When I reached Bohr he had only a few minutes left; but I had hardly begun to tell him, when he struck his forehead with his hand and exclaimed: "Oh what idiots we all have been! Oh but this is wonderful! This is just as it must be! Have you and Lise Meitner written a paper about it?" I said, we hadn't yet but would at once, and Bohr promised not to talk about it before the paper was out. Then he was off to catch his boat.

The paper was composed by several long-distance telephone calls, Lise Meitner having returned to Stockholm in the meantime. I asked an American biologist who was working with Hevesy what they call the process by which bacteria divide; "fission", he said, and so I used the term "nuclear fission" in that paper. Placzek was sceptical; couldn't I do

10

some experiments to show the existence of those fast-moving fragments of the uranium nucleus? Oddly enough, that thought hadn't occurred to me, but now I quickly set to work, and the experiment (which was really very easy) was done in two days, and a short note about it was sent off to 'Nature' together with the other note I had composed over the telephone with Lise Meitner. About six weeks went before 'Nature' printed those notes, and a good many things happened before then.

In the boat which was bringing him to America, Bohr had no relief until he had completely satisfied himself that he really understood the mechanism of the fission process. Together with Rosenfeld, who was accompanying him (and from whom I have the following details), he examined the problem from every possible angle, in his usual way, and eventually found a very simple argument showing why the fission of the nucleus, as soon as the forces tending to prevent it are sufficiently weak, will have a fair chance to occur, even in competition with the other possible and more familiar types of disintegration. In New York, Wheeler was awaiting the arrival of the boat, to accompany Bohr and Rosenfeld to Princeton; but Bohr, who had some business in New York, was left behind: he would only proceed to Princeton a couple of days later. Before his arrival, however, Rosenfeld had been invited to attend the usual meeting of the "journal club" at the Palmer Laboratory, and had been politely asked whether there was any news he could tell about. Now, during all the discussions on the boat, Bohr had not mentioned his promise to keep the fission problem to himself until our note was out; Rosenfeld was in fact under the impression that the note had just appeared or was immediately forthcoming. He therefore had no inhibition to tell the audience the whole thing, including the argument Bohr had just found, which clinched the matter in the most convincing way. The communication was recieved, as one can imagine, with considerable excitement.

Bohr was annoyed when he heard about this premature disclosure, and knowing our leisurely ways at the Institute, he tried to convey to us a sense of the urgency of the matter by sending us telegram after telegram, asking for further information and suggesting further experiments. Some of these we managed to perform, but we had no idea of the reasons which could prompt Bohr to such an unusual impatience. In fact, Bohr acted only on a suspicion of what could possibly happen, but he at first did not know that a fantastic race was already going in a number of American laboratories, where the news had spread from Princeton, to perform the same easy experiments I had already made to detect the fission fragments; a meeting of the American Physical Society would soon take place in Washington, and they all wanted to have sensational findings to report. The whole extent of the affair came out at this meeting, and for some time afterwards, Bohr and Rosenfeld (the latter, as he told me, much dismayed by the unexpected consequences of his well-meaning but imprudent communicativeness) had a rough job establishing the true priority.

One incident will illustrate what heights the excitement reached. A group of local physicists, who had not been tipped beforehand and only heard of the matter at the meeting, rushed to their laboratory and worked without pause for two days in order to be able to announce at the meeting that they too had seen the fission fragments. They invited Bohr to have a look at them, and Rosenfeld described to me the scene Bohr and he witnessed: a physicist was simultaneously looking at the apparatus recording the fragments as they were registered, and phoning to an anxious newspaper man: "Now, there's another one ..." There was a report the next day in the Washington "Evening Star". By that time, however, Bohr had heard about my own experiments, not from me (I am ashamed to confess), but from his son Hans. This was the source of the story—reprinted many times—that I was Bohr's son-in-law (although he had no daughter, and I was then unmarried). I can see how it happened: a journalist asks: "How do you know of this, Dr. Bohr?" Bohr: "My son wrote to me". Journalist mutters: "His son, but name is Frisch; must be son-in-law."

While this turmoil was taking place in the USA, we were quietly pursuing our work in Copenhagen. Lise Meitner felt that probably most of the radioactive substances which had been thought to lie beyond uranium were also fission products; a month or two later she came to Copenhagen and we proved that point by using a technique of 'radioactive recoil' which she had been the first to use, about thirty years previously. And yet in all this excitement we had missed the most important point: the chain reaction.

I think it was several weeks later that Møller first suggested that the fission fragments might contain enough energy to send out a neutron (or even two). My immediate answer was that in that case no uranium ore

10*

deposits could exist: they would have blown up long ago by the explosive multiplication of neutrons in them! But I quickly saw that my argument was too naive; ores contained lots of other elements which might swallow up the neutrons; and the seams were perhaps thin, and then most of the neutrons would escape. With that, the exciting vision arose that by assembling enough pure uranium (with appropriate care!) one might start a controlled chain reaction and liberate nuclear energy on a scale that mattered. Of course the spectre of a bomb was there as well; but for a while, anyhow, it looked as though it need not cause us much fear. That complacency was based on an argument by Bohr, which was subtle and appeared quite sound.

Bohr, in a paper on the theory of fission that he wrote with John Wheeler, had deduced that most of the neutrons emitted by the fission fragments would be too slow to cause fission of the chief isotope, U^{238}. The observed fission by slow neutrons he attributed to the rare isotope U^{235}, and he concluded that the only chance of getting a chain reaction with natural uranium was to arrange for the neutrons to be slowed down, whereby their effect on U^{235} is increased. But in that manner one could not get an efficient explosion; slow neutrons take their time, and even if one did get the conditions for rapid neutron multiplication this would at best (or at worst!) cause the assembly to disperse itself, with only a minute fraction of its nuclear energy liberated.

All this was quite correct, and the development of nuclear reactors followed closely the lines which Bohr foresaw within a few months of the discovery of fission. What he did not foresee was the fanatical ingenuity of the allied physicists and engineers, driven by the fear that Hitler might develop the decisive weapon before they did. I was in England when the war broke out, and in Los Alamos when I saw Bohr again. By that time it was clear that there were even two ways for getting an effective nuclear explosion: one through the separation of the highly fissile isotope U^{235}, the other by using the new element plutonium, formed in a nuclear reactor. Many of us were beginning to worry what the future might hold for a humanity in possession of such a dreadful weapon; and once again it was Bohr who taught us to think constructively and hopefully about that situation. But these are things about which others are better qualified than I to tell you.

"Letter to F. D. Roosevelt from Albert Einstein." From
<u>Einstein on Peace</u> edited by Otto Nathan and Heinz Norden,
Simon and Schuster, New York, 1960. Reprinted by permis-
sion of the estate of Albert Einstein. Courtesy Franklin D.
Roosevelt Library, Hyde Park, New York.

Albert Einstein
Old Grove Rd.
Nassau Point
Peconic, Long Island

August 2nd, 1939

F.D. Roosevelt,
President of the United States,
White House
Washington, D.C.

Sir:

Some recent work by E.Fermi and L. Szilard, which has been communicated to me in manuscript, leads me to expect that the element uranium may be turned into a new and important source of energy in the immediate future. Certain aspects of the situation which has arisen seem
to call for watchfulness and, if necessary, quick action on the part
of the Administration. I believe therefore that it is my duty to bring
to your attention the following facts and recommendations:

In the course of the last four months it has been made probable -
through the work of Joliot in France as well as Fermi and Szilard in
America - that it may become possible to set up a nuclear chain reaction
in a large mass of uranium,by which vast amounts of power and large quantities of new radium-like elements would be generated. Now it appears
almost certain that this could be achieved in the immediate future.

This new phenomenon would also lead to the construction of bombs,
and it is conceivable - though much less certain - that extremely powerful bombs of a new type may thus be constructed. A single bomb of this
type, carried by boat and exploded in a port, might very well destroy
the whole port together with some of the surrounding territory. However,
such bombs might very well prove to be too heavy for transportation by
air.

-2-

The United States has only very poor ores of uranium in moderate quantities. There is some good ore in Canada and the former Czechoslovakia, while the most important source of uranium is Belgian Congo.

In view of this situation you may think it desirable to have some permanent contact maintained between the Administration and the group of physicists working on chain reactions in America. One possible way of achieving this might be for you to entrust with this task a person who has your confidence and who could perhaps serve in an inofficial capacity. His task might comprise the following:

a) to approach Government Departments, keep them informed of the further development, and put forward recommendations for Government action, giving particular attention to the problem of securing a supply of uranium ore for the United States;

b) to speed up the experimental work,which is at present being carried on within the limits of the budgets of University laboratories, by providing funds, if such funds be required, through his contacts with private persons who are willing to make contributions for this cause, and perhaps also by obtaining the co-operation of industrial laboratories which have the necessary equipment.

I understand that Germany has actually stopped the sale of uranium from the Czechoslovakian mines which she has taken over. That she should have taken such early action might perhaps be understood on the ground that the son of the German Under-Secretary of State, von Weizsäcker, is attached to the Kaiser-Wilhelm-Institut in Berlin where some of the American work on uranium is now being repeated.

Yours very truly,

A. Einstein

(Albert Einstein)

"Letter to Albert Einstein from Franklin D. Roosevelt (unsigned)." Reprinted with permission of the Franklin D. Roosevelt Library, Hyde Park, New York.

October 19, 1939

My dear Professor:

I want to thank you for your recent letter and the most interesting and important enclosure.

I found this data of such import that I have convened a Board consisting of the head of the Bureau of Standards and a chosen representative of the Army and Navy to thoroughly investigate the possibilities of your suggestion regarding the element of uranium.

I am glad to say that Dr. Sachs will cooperate and work with this Committee and I feel this is the most practical and effective method of dealing with the subject.

Please accept my sincere thanks.

Very sincerely yours,

Dr. Albert Einstein,
Old Grove Road,
Nassau Point,
Peconic, Long Island,
New York.

"Alamogordo: An Eyewitness Account," by Brigadier General Thomas F. Farrell. From the Memorandum for the Secretary of War from Major General L. R. Groves, <u>Subject:</u> <u>The Test</u>, dated July 18, 1945. Manhattan Engineer District records, National Archives and Records Service, Washington, D. C.

Alamogordo: An Eyewitness Account

The scene inside the shelter was dramatic beyond words. In and around the shelter were some twenty-odd people concerned with last-minute arrangements prior to firing the shot. Included were: Dr. Oppenheimer, the Director, who had borne the great scientific burden of developing the weapon from the raw materials made in Tennessee and Washington, and a dozen of his key assistants--Dr. Kistiakowsky, who had developed the highly special explosives; Dr. Bainbridge, who supervised all the detailed arrangements for the test; Dr. Hubbard, the weather expert, and several others. Besides these, there were a handful of soldiers, two or three Army officers and one Naval officer. The shelter was cluttered with a great variety of instruments and radios.

For some hectic two hours preceding the blast, General Groves stayed with the Director, walking with him and steadying his tense excitement. Every time the Director would be about to explode because of some untoward happening, General Groves would take him off and walk with him in the rain, counseling with him and reassuring him that everything would be all right. At twenty minutes before zero hour, General Groves left for his station at the base camp, first, because it provided a better observation point, and second, because of our rule that he and I must not be together in situations where there is an element of danger, which existed at both points.

Just after General Groves left, announcements began to be broadcast of the interval remaining before the blast. They were sent by radio to the other groups participating in and observing the test. As the time intervals grew smaller and changed from minutes to seconds, the tension increased by leaps and bounds. Everyone in that room knew the awful potentialities of the thing that they thought was about to happen. The scientists felt that their figuring must be right and that the bomb had to go off but there was in everyone's mind a strong measure of doubt. The feeling of many could be expressed by "Lord, I believe; help Thou mine unbelief." We were reaching into the unknown and we did not know what might come of it. It can be safely said that most of those present--Christian, Jew, and atheist--were praying and praying harder than they had ever prayed before. If the shot were successful, it would be a justification of the several years of intensive effort of tens of thousands of people--statesmen, scientists, engineers, manufacturers, soldiers, and many others in every walk of life.

In that brief instant in the remote New Mexico desert the tremendous effort of the brains and brawn of all these people came suddenly and startlingly to the fullest fruition. Dr. Oppenheimer, on whom had rested a very heavy burden, grew tenser as the last seconds ticked off. He scarcely breathed. He held on to a post to steady himself. For the last few seconds, he stared directly ahead and then when the announcer shouted "Now!" and there came this tremendous burst of light followed shortly thereafter by the deep growling roar of the explosion, his face relaxed into an expression of tremendous relief. Several of the observers standing back of the shelter to watch the lighting effects were knocked flat by the blast.

The tension in the room let up and all started congratulating each other. Everyone sensed, "This is it!" No matter what might happen now, all knew that the impossible scientific job had been done. Atomic fission would no longer be hidden in the cloisters of the theoretical physicists' dreams. It was almost full grown at birth. It was a new force to be used for good or for evil. There was a feeling in that shelter that those concerned with its nativity should dedicate their lives to the mission that it would always be used for good and never for evil.

Dr. Kistiakowsky, the impulsive Russian (interpolation by Groves at this point: "an American and Harvard professor for many years"), threw his arms around Dr. Oppenheimer and embraced him with shouts of glee. Others were equally enthusiastic. All the pent-up emotions were released in those few minutes and all seemed to sense immediately that the explosion had far exceeded the most optimistic expectations and wildest hopes of the scientists. All seemed to feel that they had been present at the birth of a new age--The Age of Atomic Energy--and felt their profound responsibility to help in guiding into right channels the tremendous forces which had been unlocked for the first time in history.

As to the present war, there was a feeling that no matter what else might happen, we now had the means to insure its speedy conclusion and save thousands of American lives. As to the future, there had been brought into being something big and something new that would prove to be immeasurably more important than the discovery of electricity or any of the other great discoveries which have so affected our existence.

The effects could well be called unprecedented, magnificent, beautiful, stupendous and terrifying. No man-made phenomenon of such tremendous power had ever occurred before. The lighting effects beggared description. The whole country was lighted by a searing light with the intensity many times that of the midday sun. It was golden, purple, violet, gray and blue. It lighted every peak, crevasse and ridge of the nearby mountain range with a clarity and beauty that cannot be described but must be seen to be imagined. It was that beauty the great poets dream about but describe most poorly and inadequately. Thirty seconds after the explosion came the air blast, pressing hard against the people and things, to be followed almost immediately by the strong sustained, awesome roar which warned of doomsday and made us feel that we puny things were blasphemous to dare tamper with forces heretofore reserved to the Almighty. Words are inadequate tools for the job of acquainting those not present with the physical, mental and psychological effects. It had to be witnessed to be realized.

I

Atomic Explosives

by

J. ROBERT OPPENHEIMER

> THIS LECTURE *was delivered in Pittsburgh, Pennsylvania, before the George Westinghouse Centennial Forum on May 16, 1946.*

THIS TALK is to be a brief report on the future of atomic explosives. It will have to be a very incomplete and a very one-sided talk; I can hope that you will agree with me that the part of the matter that I can discuss, if not the most entertaining, is at least the most important.

When I looked over my notes for this talk, I was reminded of a story, very old and not very funny, but relevant. There was a professor of zoology at the University of Munich, and he had the habit of asking candidates about worms, until it came to such a pass that candidates studied no other subject but worms. And then, one day, he flabbergasted his student and said, "Tell me about elephants," and the candidate said, "The elephant is a large animal. It has a wormlike trunk. Worms may be divided into the following classes . . ." My talk to you this afternoon will be along these lines.

I cannot tell you of the probable future technical developments of atomic explosives. When the war was over we recognized that we had only scratched the surface of this problem; and no doubt since then some further progress has been made, both in development and in understanding. But these are things that we cannot talk about here. When, if ever, they can be talked about openly, it will be a very different world, and to my way of thinking a very much better one.

As for the uses of atomic explosives, the one that has been most widely discussed, the one in which their pre-eminence was first established and is most obvious, is the strategic bombardment of cities. No doubt there can be important tactical applications as well. I have even heard some discussion of the possibility of using them against naval craft, but on these ignorance and inexperience, as well as the requirements of secrecy, keep me from talking. There has even been a little talk of possible beneficent applications of atomic explosives, such as the blasting of polar ice or the possible control of major natural phenomena such as tornadoes, earthquakes, eruptions. There is enough energy in atomic explosives to give these vague suggestions an air of plausibility; even the weapons so far used release an energy about one thousandth of that in the San Francisco earthquake. But of course the forces produced by an atomic explosion have a very different sort of order from those involved in the great natural phenomena of quakes and of tornadoes; and the radiation and radioactivities that accompany any major atomic explosion must at least complicate its application to benign purposes. If men are ever to speak of the benefits of atomic energy, I think these applications will at most play a very small part in what they have in mind.

There is only one future of atomic explosives that I can regard with any enthusiasm: that they should never be used in war. Since in any major total war, such as we have lived through in these late years, they will most certainly be used, there is nothing modest in this hope for the future: It is that there be no such wars again. I should like to speak today on some considerations bearing upon the realization of that hope. This is a subject that seems to me worthy of careful study, and of the best thought of our times.

Some months ago, I had the privilege of working with a group of consultants to the Secretary of State's Committee on Atomic Energy. We spent many weeks exploring this problem, which is commonly defined in a sort of code as "The International Control of Atomic Energy." This is a code because the real problem is the prevention of war. Since that time our conclusions, expurgated of all secret or classified matter, have been made public, and may in one way or another have come to your attention. They were made public in order to facilitate public understanding and discussion, a discussion made more necessary by the difficulty of the problem, made more difficult by the secrecy which has been maintained and is still maintained about many of its technical elements. What I should like to do today is to add a few comments which may help to supplement the report that was made public, and to make explicit some of the things left implicit in it, to restore a balance of emphasis which was partially lost, perhaps, in the accidents of its release.

The heart of our proposal was the recommendation of an International Atomic Development Authority, entrusted with the research, development, and exploitation of the peaceful applications of atomic energy, with the elimination from national armaments of atomic weapons, and with the studies and researches and controls that must be directed toward that end. In this proposal we attempted to meet, and to put into a constructive context, two sets of facts, both long recognized, and commonly regarded as contributing to the difficulty, if not to the insolubility, of the problem.

The first of these facts is that the science, the technology, the industrial development involved in the so-called beneficial uses of atomic energy appear to be inextricably intertwined with those involved in making atomic weapons. You will hear reports this afternoon on the so-called beneficial uses of atomic energy. They come to us not in the form of answers but in the form of questions, and that for two reasons: In the first place, one of these uses is for the development of power, and this is something that has not been effectively done. No one knows to what extent such power will be economically profitable; no one knows to what extent technical problems may delay or complicate the development of atomic power as power. We have here a beginning; but we don't have any answers. We don't have a tree with fruit ripe on it, for us to shake the fruit down. The other application is in essence to research; and it is in the nature of research that you pay your "two bits" first, that you go in and you don't know what you are going to see. Therefore, if I speak of "beneficial applications," I want to make it clear that I don't know at all precisely what they are, but I share the belief that is widespread in the American people that a development of this kind in the hands of intelligent and resourceful men will lead to good things. The beginnings of these things you will hear described today.

But one thing I must go into: The same raw material, uranium, is needed for the use of atomic energy for power as for atomic bombs. The plants of an atomic power program may not be ideally suited for the production of bomb materials, but in a pinch—and atomic warfare is a pinch—they can be made to do. The various fissionable materials derived from uranium and thorium that play such a decisive part in the power program, or even in the use of atomic energy for research reactors and for advancing science and medicine and the practical arts, are, or can with more or less effort, be made into atomic explosives. The same physics which must be learned and studied and extended in the one field will help with the other—although there are of course some things in the higher art of bomb-making that as yet appear to have no other application. It is true that the properties that make a fissionable material, that make it useful for reactors for power or for research, are not quite the same properties that make it useful for bombs. Natural uranium can be used in a power plant, but I don't think a bomb can be made of it. Uranium considerably enriched in the isotope 235 can be more flexibly, more effectively used in a reactor; but I am not sure that it can be made explosive, and am fairly confident that it would be so ineffective as not to warrant the effort. Even plutonium can be doctored—not without prohibitive cost if it is to be completely nonexplosive—but to be made a relatively very ineffective explosive, and a difficult one to use in the present state of the art. I don't need to tell you that the art may change, and that no kind of control is worth anything which doesn't make provision for such change. It's not only that it can; it probably will, in one way or another. These differences in the requirements for controlled and explosive uses of atomic energy, might, if appropriately recognized in law, keep a group of individuals from making atomic weapons out of the materials of peacetime industry; they could retard and thus perhaps discourage nations otherwise prevented from the exploitation of atomic energy; but this isn't the problem; for to any who are actively engaged in such exploitation they could provide a deterrent so slight as to constitute a most dangerous illusion. Thus a mere prohibition on the activities of nations in the field of atomic energy sufficiently incisive to inspire confidence that, if enforced, it would prevent rapid conversion to atomic armament, would at the same time close this field to the exploitation of any of its benefits. This fact, which further technical developments appear unlikely to invalidate, has long been regarded as an almost decisive difficulty on the path of international control. It might have appeared so to us, too, if there had not been a greater one. For even if the course of development of atomic energy for peace were entirely distinct from its development for war, even if it were universally agreed that there were no peaceful applications of atomic energy worthy of interest or of effort, we should still be faced with the fact that there exists in the world today no machinery for making effective a prohibition against the national development of atomic armaments. In the light of this fact, that to my mind touches upon the heart of the problem, the close technical parallelism and interrelation of the peaceful and the military applications of atomic energy ceases to be a difficulty, and becomes a help. This does not, unfortunately, mean that it guarantees a solution. But it does mean that it provides a basis for seeking a healthy solution that would not otherwise exist.

If there were nothing to do with atomic energy but make bombs, there might still, it is true, be a convention between nations not to do so. Such conventions have in the past seldom withstood the strain of rivalries between nations preparing for war, nor does it seem likely that they could do so in the future in the case of a weapon whose effectiveness, especially in surprise, is so spectacular. For this reason two proposals have long been current for supplementing international conventions with some form of international action. One of these would set up a scheme of multilateral or international inspection, whose sole function would be to attempt to establish that the conventions were in fact being observed. It is conceivable that if the conventions were sufficiently radical, comprising, for instance, the total renunciation of all mining and refining of uranium, such a procedure might work. But I doubt this, even in that case. I doubt whether the agency entrusted with such inspection could even then have the motivation, or the personnel, or the skill, or the experience, or the knowledge, or the endurance to carry out such a dreary, sterile, and policeman-like job. I doubt whether the relations between this agency and the nations and nationals whom it was instructed to police would be such as to diminish the nationalism leading to war, or to inspire the confidence of the nations in each other, or to advance the cause of the unification of the world, or to serve as a useful prototype for the elimination of weapons of mass destruction, perhaps equally, perhaps even more terrible. Therefore one may perhaps not regret that the door to this sort of international action is largely closed by the impossibility of denying to the world in any long term an opportunity to explore the beneficial possibilities of atomic energy. For once such exploration is allowed to the nations, the technical complexities and human inadequacies of an international inspection scheme as a sole safeguard become manifestly insupportable.

The second suggestion for international action to supplement the renunciation by nations of atomic armaments has a more affirmative character. It is that an international agency be entrusted with the making and possession of atomic weapons. Though there has been much in this proposal that has seemed attractive, it has two weaknesses, and probably fatal ones: The more serious is that there is nothing that an international agency can do, or should do, with such weapons. They are not police weapons. They are singularly unsuited for distinguishing between innocent and guilty or for taking even crudely into account the distinction between the guilt of individuals and that of peoples; they are themselves a supreme expression in a weapon of the concepts of total war. The second difficulty, in some sense inescapable in any form of international action, but desperately acute in this, is that such stocks of atomic weapons, however earnestly they are proclaimed international, however ingeniously they are distributed on earth, would nonetheless offer the most terrible temptation to national seizure, for the almost immediate military advantage that their use might afford.

These two examples do give recognition to the need, in any system of outlawing atomic weapons, of international action. In this I think they are sound. In fact, in another context, the study—but not the production—of atomic weapons, and inspection to prevent the illegal mining of uranium, both would seem to be essential functions of an international authority.

It is time to turn to the second of the great difficulties that have from the outset been regarded as preventing any effective international control. We have already referred to it. It is the absence in the world today of any machinery adequate to provide such control, any precedent for such machinery, or even any adequate patterns of the past to provide such a precedent. Just this is the reason why the problem is so much of a challenge, why we may be sustained by the hope that its solution would provide such precedent, such patterns, for a wider application. It did not take atomic weapons to make wars, or to make wars terrible, or to make wars total. If there had never been and could never be an atomic bomb, the problem of preventing war in an age when science and technology have made it too destructive, and too terrible to endure, would still be with us. There would be the block-buster, the rocket, the V-2, the incendiary, the M-67, and their increase; there would no doubt be biological warfare. There would be, and there still are. But the atomic bomb, most spectacular of proven weapons, the most inextricably intertwined with constructive developments and the least fettered by private or by vested interest or by long national tradition is for these and other reasons the place to start. For in this field there is possible a system of control that is consistent with, that is based upon, the technical realities and with the human realities in the deep sense. In this field, there is a solution that can be made to work.

Many have said that without world government there could be no permanent peace, and without peace there would be atomic warfare. I think one must agree with this. Many have said that there could be no outlawry of weapons and no prevention of war unless international law could apply to the citizens of nations, as federal law does to citizens of states, or have made manifest the fact that international control is not compatible with absolute national sovereignty. I think one must agree with this. Many have said that atomic energy could not be controlled if the controlling authority could be halted by a veto, as in many actions can the Security Council of the United Nations. I think one must agree with this too. With those who argue that it would be desirable to have world government, an appropriate delegation of national sovereignty, laws applicable to individuals in all nations, it would seem most difficult to differ; but with those who argue that these things are directly possible, in their full and ultimately necessary scope, it may be rather difficult for me to agree.

What relation does the proposal of an International Atomic Development Authority, entrusted with a far-reaching monopoly of atomic energy—what relation does this proposal of ours have to do with these questions? It proposes that in the field of atomic energy there be set up a world government, that in this field there be renunciation of national sovereignty, that in this field there be no legal veto power, that in this field there be international law. How is this possible, in a world of sovereign nations? There are only two ways in which this ever can be possible: one is conquest, that destroys sovereignty; and the other is the partial renunciation of that sovereignty. What is here proposed is such a partial renunciation, sufficient, but not more than sufficient, for an Atomic Development Authority to come into being, to exercise its functions of development, exploitation and control, to enable it to live and grow and to protect the world against the use of atomic weapons and provide it with the benefits of atomic energy.

Whatever else happens, there is likely to be a discussion of the control of atomic energy in the United Nations Commission set up for that purpose, and not in the very distant future, I would say. Should these discussions eventuate in the proposal of an International Authority, and in a charter for that Authority, these proposals and that charter would in the end be presented for ratification to the several nations. Each nation, the small as well as the great, can exercise its sovereign right to refuse such ratification. Should that happen, there would be no Atomic Development Authority, and in my opinion probably no trustworthy, effective, international control of atomic energy. Should a nation, after the creation of the Authority, exercise its sovereign right and withdraw from it, or fail with regard to it to carry out the accepted and major conditions of the charter, then there will also be no Atomic Development Authority; unlike the Security Council, it presumably could not survive the application of the veto to its major provisions. But if it comes into existence, and insofar as it stays in existence, it will provide, in this field, the international sovereignty whose necessity has been so generally recognized.

Perhaps, one will say, no international enterprise can live under such conditions. But the conditions themselves will not remain unaffected by the enterprise. Its coming into existence will be a step that, once learned, can be repeated, a commitment that, once made in one field, can be extended to others. If this is to happen, the Development Authority will have to have a healthy life of its own; it will have to flourish, to be technically strong, to be useful to mankind, to have a staff and an organization and a way of life in which there is some pride, and some cause for pride. This would not be possible if there were nothing of value to do with atomic energy. This would not be possible if the prevention of atomic armament were its only concern, if all other activity was technically so separable and separate from atomic armament that it could remain in national hands. In the long struggle to find a way of reconciling national and international sovereignty, the peaceful applications of atomic energy can only be a help. It is perhaps doubtful that we should have had a federal government had not those functions which could not safely nor effectively be carried out by the states had a certain importance for the people of this country.

The Board of Consultants to the State Department was aware of the supreme necessity for providing the Authority with work that could attract men, and consolidate and inspire them. It was equally aware of the complementary dangers of a too complete, a too absolute monopoly. These dangers are of two kinds: on the one hand, a monopoly which is not subject to criticism is likely to go to seed; it is likely not to be on its toes; it is likely in the end to become bureaucratically inbred. On the other hand, if you have no living, legitimate contact between the operations of an Authority like this, and the activities of scientists, engineers, and business men operating outside the Authority, in national or in private agencies, then you have no way of being sure that you are not missing many important bets. A too absolute monopoly would be dangerous both to the health of the monopoly and to the surveillance activities which an Authority of this kind must maintain.

For this reason we found it important to point out that there were many activities in the field of atomic energy which either in themselves or because they are easy and reliable to control and inspect and supervise, could not lend themselves to evasive or diversionary developments of atomic weapons. An example of this kind is the whole field of the use of tracers. An example of this kind is the use of reactors for research. An example of this kind which is somewhat more marginal, is the use of reactors which burn and do not produce explosive material for power, and in which the best steps you can take to complicate and delay the use of this material for explosives have been taken, so that it isn't a thing that can be done in an hour's effort or in a month's effort or by a few angry individuals. I think the importance of this point is this: there are safe activities which you can leave, for instance, in the hands of the government of the United States or the corporations of the United States or the universities of the United States. For this reason, there will be good, technical liaison between the Authority and these more private agencies. This will, on the one hand, tend to correct the bureaucracy that is implicit in monopoly. On the other hand, it will give the International Authority some method of remaining cognizant of the developments in the field which happen not to have been carried out by itself.

If any great note of confidence or gayety has invested these brief words, it would be a distortion of the spirit in which I should have wished to speak to you. No thoughtful man can look to the future with any complete assurance that the world will not again be ravaged by war, by a total war in which atomic weapons contribute their part to the ultimate wreck and attrition of this our Western civilization. My own view is that the development of these weapons can make, if wisely handled, the problem of preventing war, not more hopeless, but more hopeful, than it would otherwise have been, and that this is so not merely because it intensifies the urgency of our hopes, but because it provides new and healthy avenues of approach. In developing these avenues the fact that there is so far-reaching a technical inseparability of the constructive uses of atomic energy from the destructive ones—a fact that at first sight might appear to render the problem only more difficult—this fact is precisely the central vital element that can make effective action possible. If we are clear on this, we shall have some guide for the future.

THE THREAT OF NUCLEAR WAR

Three technical aspects of thermonuclear weapons have made thermonuclear war a peril to the very existence of humanity. These aspects are: the enormous destructive power of a thermonuclear explosion, the relative cheapness of rocket-thermonuclear weapons, and the practical impossibility of an effective defense against a massive rocket-nuclear attack.

1

Today one can consider a three-megaton nuclear warhead as "typical" (this is somewhere between the warhead of a Minuteman and of a Titan II). The area of fires from the explosion of such a warhead is 150 times greater than from the Hiroshima bomb, and the area of destruction is 30 times greater. The detonation of such a warhead over a city would create a 100-square-kilometer [40 square-mile] area of total destruction and fire.

Tens of millions of square meters of living space would be destroyed. No fewer than a million people would perish under the ruins of buildings, from fire and radiation, suffocate in the dust and smoke or die in shelters buried under debris. In the event of a ground-level explosion, the fallout of radioactive dust would create a danger of fatal exposure in an area of tens of thousands of square kilometers.

2

A few words about the cost and the possible number of explosions.

After the stage of research and development has been passed, mass production of thermonuclear weapons and carrier rockets is no more complex and expensive than, for example, the production of military aircraft, which were produced by the tens of thousands during the war.

The annual production of plutonium in the world now is in the tens of thousands of tons. If one assumes that half this output goes for military purposes and that an average of several kilograms of plutonium goes into one warhead, then enough warheads have already been accumulated to destroy mankind many times over.

3

The third aspect of thermonuclear peril (along with the power and cheapness of warheads) is what we term the practical impossibility of preventing a massive rocket attack. This situation is well known to specialists. In the popular scientific literature, for example, one can read this in an article by Richard L. Garwin and Hans A. Bethe in the *Scientific American* of March, 1968.

The technology and tactics of attack have now far surpassed the technology of defense despite the development of highly maneuverable and powerful antimissiles with nuclear warheads and despite other technical ideas, such as the use of laser rays and so forth.

Improvements in the resistance of warheads to shock waves and to the radiation effects of neutron and X-ray exposure, the possibility of mass use of relatively light and inexpensive decoys that are virtually indistinguishable from warheads and exhaust the capabilities of an antimissile defense system, a perfection of tactics of massed and concentrated attacks, in time and space, that overstrain the defense detection centers, the use of orbital and fractional-orbital attacks, the use of active and passive jamming, and other methods not disclosed in the press—all this has created technical and economic obstacles to an effective missile defense that, at the present time, are virtually insurmountable.

The experience of past wars shows that the first use of a new technical or tactical method of attack is usually highly effective even if a simple antidote can soon be developed. But in a thermonuclear war the first blow may be the decisive one and render null and void years of work and billions spent on creation of an antimissile system.

An exception to this would be the case of a great technical and economic difference in the potentials of two enemies. In such a case, the stronger side, creating an antimissile defense system with a multiple reserve, would face the temptation of ending the dangerous and unstable balance once and for all by embarking on a pre-emptive adventure, expending part of its attack potential on destruction of most of the enemy's launching bases and counting on impunity for the last stage of escalation, i.e., the destruction of the cities and industry of the enemy.

Fortunately for the stability of the world, the difference between the technical-economic potentials of the Soviet Union and the United States is not so great that one of the sides could undertake a "preventive aggression" without an almost inevitable risk of a destructive retaliatory blow. This situation would not be changed by a broadening of the arms race through the development of antimissile defenses.

In the opinion of many people, an opinion shared by the author, a diplomatic formulation of this mutually comprehended situation, for example, in the form of a moratorium on the construction of antimissile systems, would be a useful demonstration of a desire of the Soviet Union and the United States to preserve the status quo and not to widen the arms race for senselessly expensive antimissile systems. It would be a demonstration of a desire to cooperate, not to fight.

A thermonuclear war cannot be considered a continuation of politics by other means (according to the formula of Clausewitz). It would be a means of universal suicide.

Two kinds of attempts are being made to portray thermonuclear war as an "ordinary" political act in the eyes of public opinion. One is the concept of the "paper tiger," the concept of the irresponsible Maoist adventurists. The other is the strategic doctrine of escalation, worked out by scientific and militarist circles in the United States. Without minimizing the seriousness of the challenge inherent in that doctrine, we will just note that the political strategy of peaceful coexistence is an effective counterweight to the doctrine.

A complete destruction of cities, industry, transport, and systems of education, a poisoning of fields, water, and air by radioactivity, a physical destruction of the larger part of mankind, poverty, barbarism, a return to savagery, and a genetic degeneracy of the survivors under the impact of radiation, a destruction of the material and information basis of civilization—this is a measure of the peril that threatens the world as a result of the estrangement of the world's two superpowers.

Every rational creature, finding itself on the brink of a disaster, first tries to get away from the brink and only then does it think about the satisfaction of its other needs. If mankind is to get away from the brink, it must overcome its divisions.

A vital step would be a review of the traditional method of international affairs, which may be termed "empirical-competitive." In the simplest definition, this is a method aiming at maximum improvement of one's position everywhere possible and, simultaneously, a method of causing maximum unpleasantness to opposing forces without consideration of common welfare and common interests.[4]

If politics were a game of two gamblers, then this would be the only possible method. But where does such a method lead in the present unprecedented situation?

CHAPTER XVII

Safety of Nuclear Reactors

AT THE BEGINNING of the scientific and industrial revolution two old ambitions were found to be impossible dreams. One was the transmutation of elements, the other the machine of perpetual motion.

Modern nuclear physicists had to retract one of these statements: elements can be transmuted. But the product is expensive, for the time being much more expensive than gold.

The perpetual motion machine remains impossible in principle but the problem may be considered solved in practice. It can be proved, of course, that a machine can do useful work only if it burns up some fuel. But the price of fuel is quite often less than the cost to operate and maintain the machine.

Nuclear fuel even today is no more expensive than conventional fuel in many parts of the United States. Nuclear fuel is neither heavy nor bulky and can be therefore transported easily. In those parts of the world where ordinary fuel is expensive, nuclear energy will soon become of great importance. Furthermore, we shall learn to use most of the energy in uranium rather than just the part contained in its rare and valuable isotope, U^{235}.

152

One only has to add a neutron to common U^{238} to get radioactive U^{239}. In the course of time this decays into plutonium. This element can be used like U^{235}: It produces fission, a great amount of energy and enough neutrons to keep the process going. We shall also learn to extract energy from other nuclear fuels. Thorium acts like uranium, while deuterium can give energy by building up bigger nuclei rather than breaking them into smaller pieces. Therefore the source of energy will be universally available and quite inexpensive. This really means that we are as well off as though we had a machine of perpetual motion.

But, of course, all this does not mean that the machine will do its job free of charge. Even a perpetual motion machine would need servicing and maintenance. Unfortunately our nuclear machines need a lot of such servicing and therefore for the time being, nuclear energy is not the cheapest.

The main reason why a nuclear energy source, or a nuclear reactor is difficult and expensive to run is that the reactor after a short time of operation becomes strongly radioactive. Therefore it cannot be approached and it has to be handled by remote control. We can hardly expect that energy will be free like air or water. But when we learn how to handle inexpensively our nuclear machines, we shall be able to obtain energy for a reasonable price at any place on the earth. Sooner or later conventional fuel will become scarce. But nuclear energy will allow the industrial revolution to continue and to expand into every corner of the earth.

There can be little doubt that during the next decades nuclear reactors will greatly multiply and by the beginning of the next century they will be found everywhere. It is therefore of the greatest importance that these reactors should be operated safely. On the face of it, a nuclear reactor is a sluggish instrument which can be made to run itself. But the ease of operation is deceptive. (See picture 13.)

One need not fear that a nuclear reactor might explode

like an atomic bomb. Nuclear explosives are very carefully constructed so that they can release a lot of energy in a short time. Nuclear reactors on the other hand are put together so as to make it possible that energy will be released only at a moderate rate. Some reactors if improperly handled may explode, but the violence of the explosion cannot greatly exceed that of a similar weight of high explosive.

Nevertheless a reactor accident could become exceedingly dangerous. The reactor is charged with radioactive fission products and some other radioactive substances produced by neutron absorption. Any accident which will allow even a portion of these products to escape into the air will endanger people at a considerable distance in the downwind direction. One reason why reactors can be dangerous is that in protracted operation of the reactor, fission products which have longer lives accumulate. It is precisely these longer-lived products which are more dangerous because they have a better chance to find their way into the human body.

Reactors are now planned which will produce 300,000 kilowatts of electricity. If such a reactor operates for half a year and then explodes and releases its radioactive content into the atmosphere, its radioactivity will be comparable to that of a hydrogen bomb. In one important respect such an accident would be worse than a hydrogen explosion. The nuclear explosive lifts most of its radioactive products to a high altitude and the poisonous activity gets dispersed and diluted before it descends. The activity from a reactor on the other hand will remain close to the ground and might endanger the lives of the people in an area of hundreds of square miles. It will contaminate an even greater territory.

In the extensive operation of many reactors in the United States no one has yet been killed by the radioactivity. This has been due to extremely careful operation and also to good luck. We must be prepared that sooner or later accidents will occur. On the other hand we must try to take sufficient precautions to avoid the kind of catastrophic accident which we have mentioned above. With great care such accidents can indeed be avoided.

In thinking of all kinds of man-made machines we find some which move fast and seem dangerous like, for instance, airplanes; others which are stationary and apparently harmless, like the bath tub. Yet more accidents happen in bath tubs than in air travel. The most dangerous element in all operations is the human element. We ourselves constitute the greatest safety hazard. This is a situation no different in nuclear technology than in any other kind of technology. What is new in nuclear technology is that a reactor is usually very safe but may become extremely dangerous when something unexpected happens to it. Also we dare not use the method of trial and error. An error in the reactor business could exact a far heavier toll of lives than an error in the testing of H-bombs. We cannot wait to learn by experience; we must forestall accidents.

An especially difficult safety problem is connected with the use of reactors in small countries. A serious accident could endanger the lives of people in adjacent countries. Thus modern technology may force cooperation across national boundaries.

There is only one way to avoid traffic accidents and that is care exercised by everyone, particularly the drivers. Similarly reactor safety will depend on the people who operate the reactors. At the same time a lot of help can be obtained by careful construction and scrutiny of each new reactor.

One of the first acts of the Atomic Energy Commission was to establish a Committee for Reactor Safeguards. With the passing of years this committee had to take on more heavy responsibilities. At first it had to operate under secrecy. With the wider and more public use of reactors the safety considerations are becoming more available to the public. The question of safe operation of a machine cannot be separated from a thorough understanding of the working of the machine. We cannot attempt to give an adequate description of a reactor or of the safety rules. A few general statements have to suffice.

A working reactor is full of neutrons. In a small fraction of a second these neutrons produce fission and a new generation of neutrons comes into being. In slow reactors which contain lots of light elements like hydrogen or carbon, the neutrons move with speeds little greater than that of sound and a generation may last as long as a millisecond (one thousandth of a second). In fast reactors which contain almost exclusively heavier elements like uranium or iron, neutrons move with a great speed which is about three per cent of the speed of light. In this case one generation replaces another in less than a microsecond (one millionth of a second).

Fortunately not all the neutrons get reproduced so rapidly. Some fissions produce delayed neutrons which are emitted usually with a delay of several seconds. In a steadily working reactor each generation should have the same number of neutrons as the previous one. If each succeeding generation has even a slight surplus, the reactor will become hot and may explode in a small fraction of a second. The main reason why safe operation is possible is the fact that fast multiplication can occur only if each generation becomes more populous *even when one does not count the delayed neutrons*. A slightly overactive reactor is easily governed, but there comes a point when the dormant dragon begins to stir. This happens when there are enough neutrons produced so that multiplication can occur without waiting for the delayed neutrons. At that point a well behaved dragon will perform a harmless action. For instance it may blow a fuse. But a vicious dragon will spit radioactive fire.

It is not easy to predict whether the dragon will be always well behaved. But with careful analysis one can make such a prediction. For instance one must look into the question of whether the reactor is stable. If it gets hotter, does this make the reactor proceed even faster so that the rate of heating increases and the reactor runs away? In a stable reactor excess heat should tend to stop the energy production and thus the reactor cools and returns to its normal operating temperature.

But too great a stability may also be dangerous. Heating may be overcompensated by the cooling mechanism; after the reactor has become too cold it may then heat up too fast and overshoot again. We must guard not only against a simple run-away, but also against increasing oscillations.

In many reactors unusual chemical compounds are used. A reactor accident may start with nothing worse than an ordinary chemical reaction between strange compounds under strange conditions. But if this chemical reaction destroys the reactor sufficiently to allow some fission products to escape, then such a chemical accident can be as bad as one of nuclear origin.

In the interior of the reactor materials are exposed to unusually strong radiation. Under this effect some materials can change their chemical properties so that what has been inert as a construction material may become dangerous during the operation of the reactor.

Perhaps the most important single item is the arrangement of mechanical controls. The reactor is adjusted by a system of sheets or rods made of a material which absorbs neutrons. This arrangement must be so constructed that the control rods can be withdrawn only at a very slow rate. But it must be possible to put them back quite fast. Any danger signal should shove the absorbers in at maximum speed. The technical expression is "scram."

The main point, however, is that all the dangers and safety devices can be studied and after careful study a nuclear accident can be avoided. Some reactors are now so thoroughly understood that they can be safely used for training of future nuclear engineers. Other reactors which are more powerful or less well studied have to be used more carefully. Some reactors should be, and are being, enclosed in gas-tight containers. If an explosion occurs the fission products will be harmlessly confined inside the container. Of course, one must be quite sure that the reactor is of such a type that it cannot produce an explosion great enough to burst the container and what is even more important one should be quite sure that the container is closed except when the reactor is shut down and completely safe. Often it may be best to build the reactor underground.

The safety of a reactor, of course, depends to a great extent on the use to which the reactor is put. In general a power station is less likely to give trouble than a moving power source. It is not probable that nuclear locomotives will ever be safe. In nuclear ships more room is available and more room permits more safety measures. But even so the safety of nuclear motors in ships will have to be considered particularly carefully because ships will have accidents in harbors.

Between the urgent need for progress and the absolute necessity of safety it is difficult to keep a sense of balance and one can easily make the mistake of being unnecessarily cautious. Such unnecessary caution was probably exercised when the Committee on Reactor Safeguards considered the earthquake hazard of the Brookhaven reactor on Long Island. A seismologist, who is a Jesuit Father, was asked to tell the committee[1] of the possibilities and probabilities of an earthquake on Long Island. The chairman[2] of the committee subjected the expert to a long and detailed questioning. After half an hour the Committee on Reactor Safeguards ran out of questions. But the Jesuit Father had not given any signs of running out of answers. The session being at an end the expert, looking the chairman of the committee firmly in the eye and

[1] Dubbed by its friends "Committee for Reactor Prevention."
[2] One of the authors.

in a more authoritative voice than he had yet used, said, "Mr. Chairman, I can assure you on the highest authority that there will be no major earthquakes on Long Island in the next fifty years."

EXCAVATION
with
NUCLEAR
EXPLOSIVES

...promise and problems

By Gerald W. Johnson

THE Plowshare Program, devoted to exploring the potential constructive uses of nuclear
explosions, was formally established in 1957
by the Atomic Energy Commission. Many of the
ideas, thoughts, and suggestions for such applications had been put forth from the time it appeared
possible to release energy from the nucleus in a
controlled manner, using either reactors or explosions. In fact, in 1939, immediately after the publication of the discovery of the fission of uranium
early in that year, many speculations appeared in
the press concerning possible industrial uses of nuclear energy. Congress, in writing the Atomic Energy Act, noted, in the declaration of that Act, that
"atomic energy is capable of application for peaceful as well as military purposes". In addition, the
stated policy of the United States is that:

"a. the development, use, and control of atomic energy
shall be directed so as to make the maximum contribution
to the general welfare, subject at all times to the para-
mount objective of making the maximum contribution to
the common defense and security; and

"b. the development, use, and control of atomic energy shall be directed so as to promote world peace, improve the general welfare, increase the standard of living, and strengthen free competition in private enterprise."

As is always the case with a new technology, much of the research and development is partially justified and funds are appropriated on the basis of projected industrial and civil use. Atomic energy has been no exception. However, a rather curious thing is happening in the United States: while national investments in research and development are increasing, the apparent return is not increasing nearly so rapidly—in fact, over the past twenty years the trend appears rather to be retrograde.

In the period 1947 to 1954, the average rate of growth of the economy was 3.7 percent; from 1954 to 1960 it was 3 percent. This took place in a period when our research and development expenditures tripled and the percentage of gross national product spent on research and development doubled (from 1.4 to 2.8 percent). Of the 2.8 percent, 2 percent is for space, defense, and atomic energy and 0.8 percent for all other purposes. Much of the justification of these enormous expenditures is based on the assumption that results will directly benefit industry and the economy. In commenting on this point recently, Mr. Holloman, Assistant Secretary of Commerce for Science and Technology,[1] had this to say: "These efforts to develop military equipment, atomic power plants, and space vehicles may well be providing the basis for a whole new technology of complex systems made up of highly reliable parts, but the translation of this technology to the economy through industry and commerce is neither direct nor cheap—nor inevitable. In fact, the translation requires specially trained people with a special point of view and an industry that understands and appreciates the possibilities of the new technology and can afford to use it. These people come from the same pool of scientists and engineers who provide the technology to meet the threat to our national security."

The Plowshare Program is an attempt to translate the technology of nuclear weaponry to the works of man. To accomplish this will require effort and will employ the same technology and people who have been responsible for the tremendous developments in weapons. While there are several goals of the program, I want to focus now on one which appears to be within reach if we want to use it—namely, the employment of nuclear explosions for excavation. My reason for selecting this area for discussion is that I believe the margin of potential cost advantage in using nuclear excavation on large projects is sufficient and that economic factors are not likely to be controversial. It also is an area in which major contributions to the public welfare can be made in the immediate future in terms of water-resource development and conservation, commerce (harbors and canals), and mining. In addition to attacking our own problems, through developing these methods and making them available on an international basis, important steps in international cooperation may also be taken.

The simple technical idea of nuclear excavation involves the detonation of nuclear charges, either singly or in an array, to provide a crater or a ditch for appropriate engineering purposes.

The principal problems appear to involve the development of techniques and the experimental demonstration that public health and safety are in fact assured so that public apprehensions centered around psychological-political factors may be overcome.

The bulk of the experimental program over the past few years has been directed toward developing suitable explosives, understanding the physical processes involved in cratering, and verifying theoretical models through full-scale experiments. Of most interest has been the determination of the sizes and characteristics of the craters formed, depending on the magnitude of the charge (scaling), the depth of placement, the nature of the medium, and, for multiple charges, the spacing between charges. Since 1957 a large part of the Plowshare Program has been devoted to carrying out experiments with charges of chemical explosives ranging from 256 pounds to 500 tons in a single event. These charges were spherical and centrally ignited to duplicate the characteristics of a nuclear explosion as closely as possible. This work established a scaling law for chemical explosives over the experimental range of yield used, which covers a factor of 4000 in energy. The derived relationships showed the crater diameter and depth to be proportioned to $W^{1/3.4}$, where W is the yield. In addition, the dependence of crater dimensions on depth of burst was well established for Nevada desert alluvium (a lightly cemented sand and gravel). The results of this work are plotted in Fig. 1 and Fig. 2 (courtesy M. Nordyke [2]), where the data are all corrected to 1 kt using the empirical scaling law.

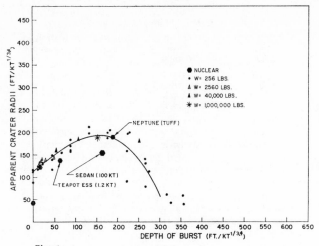

Fig. 1

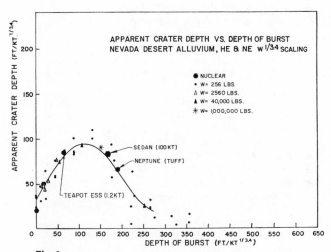

Fig. 2

Shown on the same graphs are the results obtained with nuclear explosions, which consisted of three nuclear experiments at 1.2 kt and one at 500 tons in the same medium fired for military-effects purposes (JOHNIE BOY, not plotted), one 115-ton nuclear weapons-development experiment in tuff (NEPTUNE), one 500-ton military experiment in basalt (DANNY BOY, not plotted), and one 100-kt Plowshare cratering shot. The experiments together with the results are given in Table 1. Except for the 1.2-kt event slightly above the surface

333

Table 1. Nuclear Cratering Tests

Event	Yield (kt)	Medium	Depth of burst (feet)	Crater radius (feet)	Crater depth (feet)	Crater volume (yd³)
JANGLE S	1.2	Alluvium	−3.5ᵃ	45	21	2470
JOHNIE BOY	0.5	Alluvium	2	62	30	6650
JANGLE U	1.2	Alluvium	17	130	53	3.7×10^4
TEAPOT ESS	1.2	Alluvium	67	146	90	9.6×10^4
DANNY BOY	0.42	Basalt	112	110	63	4.4×10^4
NEPTUNE	0.115	Tuff	100ᵇ	100	35	2.2×10^4
SEDAN	100.00	Alluvium	635	600	320	6.7×10^6

ᵃ The distance to the center of the explosive—above the surface.
ᵇ NEPTUNE was detonated 100 feet beneath a 30° slope.

(3.5 feet), the agreement of the scaled nuclear data with the high-explosive results is striking. If the scaling law derived from the chemical-explosive experience is used, the observed crater depths for nuclear experiments lie on the empirical curves, but the radii are fifteen to twenty percent smaller.

When most of this work was considered together with some experimental work with chemical explosives using spaced charges, it was possible to make some estimates of sizes and costs of various excavation projects. Such estimates for single craters published in 1960 [3] are listed in Table 2. The costs were based on published estimates of costs of nuclear explosives developed for military systems and on the cratering data available at that time. Both the projected costs and dimensional characteristics must now be revised in the light of new results and developments. For example, all predicted volumes must be reduced by a factor of two. Economically, this makes slight difference although it will require an approximate doubling of yields since there will be little change in cost.

As a consequence of the large strides made in the technology of nuclear explosives in the last two years, it now appears that an explosive can be provided for Plowshare purposes in the near future with the following characteristics:

1. It can be lowered down a 36-inch hole;
2. It will weigh not more than 10 000 pounds;
3. It can provide an energy release at selected values from 100 kt to 1 Mt;
4. Its release of radioactivity will be greatly reduced in comparison with earlier systems.

Larger-yield and smaller-yield systems can be provided at some adjustment in cost and dimensions. However, for the present, most actual proj-

Table 2. Estimated Costs of Crater Formation in Dry Desert Alluvium

Yield (kt)	Placement Hole Diam (in)	Placement Hole Depth (ft)	Explosive Cost ($)	Placement Cost ($)	Operations ($)	Total ($)
1	36	160	500 000	100 000	500 000	1 100 000
10	36	325	500 000	150 000	750 000	1 400 000
100	70	620	750 000	300 000	1 000 000	2 050 000
1000	70	1220	1 000 000	600 000	2 000 000	3 600 000

Crater Dimensions

Diam (ft)	Depth (ft)	Volume (yd^3)	Cost ($/yd^3)
400	90	210 000	5.25
800	175	1 600 000	0.88
1600	350	12 000 000	0.17
3200	690	96 000 000	0.04

ects that have been examined can be approached successfully using explosives in the 100 kt to 1 Mt range of energy release.

Of major projects examined, the one most extensively studied is that for a sea-level canal across the American Isthmus. This has been under study for several years by the Corps of Engineers of the US Army and the Panama Canal Company in collaboration with the Atomic Energy Commission. The House Committee on Merchant Marine and Fisheries reported to the House of Representatives on June 23, 1960, that a sea-level canal probably would not be economically feasible in the foreseeable future unless nuclear methods can be used. The committee urged that development of such methods be vigorously pursued.[4] Last year the AEC set a goal of approximately five years for the development of a nuclear-excavation technology.[5] However, before any project such as a major sea-level canal could be undertaken, many smaller projects probably would have to be carried out.

Thus, from an economic standpoint, there appears to be a clear advantage in the use of nuclear excavation techniques for appropriate projects. Using the immediate technology would require projects involving excavation of a few million yards for each single charge. With development and practical experience, the costs can be expected to be reduced to the point, perhaps, where explosions in the few-kiloton range will lead to profitable excavation. For the sea-level canal excavation, factors up to several-fold in cost advantage appear to be available. For simple jobs like river diversion, over-burden removal, or reservoir construction, the cost advantage of nuclear methods may be as high as 100-fold, depending on the size and function of the job.

While these considerations are all favorable, a large amount of work still is required to develop the detailed technology of excavation. It is clear, however, that whatever the developments show they cannot be expected to change the foregoing general conclusions. In view of this, why aren't projects under way now? The reasons have to do with radioactivity and its consequences, and with political-psychological factors.

Here, it is perhaps useful to refer to the large thermonuclear shot fired on the surface at Bikini

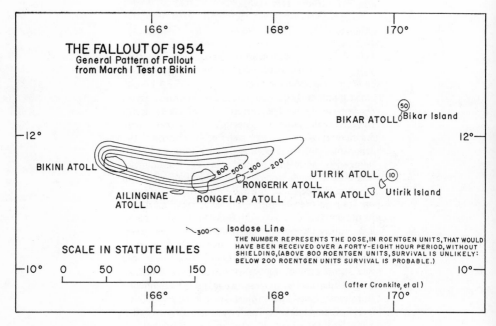

THE FALLOUT OF 1954
General Pattern of Fallout
from March I Test at Bikini

Fig. 3

on March 1, 1954. This event was a 14.5-megaton nuclear weapons-development shot fired on the surface of a coral islet at Bikini. The cloud rose to a height in excess of 100 000 feet and, instead of moving northward as expected, moved eastward, depositing large fallout on the inhabited atolls of Rongelap, Ailinginae, and Rongerik 100 miles to the east. The general levels of fallout and the pattern in terms of possible 48-hour accumulation of gamma-radiation dosage are shown in Fig. 3 (courtesy of Neal O. Hines [6]). These levels were so high that the atolls had to be evacuated. The inhabitants of Rongelap (64 men, women, and children) who were evacuated 51 hours after the detonation had received the highest calculated dose of radiation, later estimated at 175 roentgens. The other 200 people, including 28 Americans on Rongerik, received less. The sources of food in the lagoon and on the land were all heavily contaminated by the fallout. As a consequence of this unhappy event, a major ecological study was initiated at Rongelap which has continued ever since. By June 29, 1957, the levels had decayed to such a point that the native residents could be repatriated, and they have been living there ever since.

The experience in the Pacific has been recently authoritatively documented in *Proving Ground* by Neal O. Hines of the Laboratory of Radiation Biology of the University of Washington.[6] The following discussion is based on that book. During the early years the radioactivities at Rongelap continued to be redistributed within the biological pattern but had reached equilibrium by about 1960 or 1961. Henceforth it is expected that in each organism the radioactivities will decay according to their radioactive half-lives rather than changing as a consequence of redistribution due to biological processes. Various generalizations regarding radioactivities are possible but I shall limit myself to two: (1) on land, fission products appeared to be most important; and (2) in the marine environment, induced activities seemed most relevant. The experience at Rongelap and Eniwetok-Bikini is the most valuable we have in understanding the processes at work following massive fallout on biological systems involving man and his environment (and, most important, in evaluating the possible biological impact of Plowshare excavation projects). This experience is valuable because it provides upper limits of the effects of radiation on the biosphere in one type of environment. It must be noted that the use of Plowshare explosives and burial techniques is expected to result in radioactivity levels, even in the crater, much below those observed at Rongelap, and many orders below those that existed at Bikini and Eniwetok from all shots.

In the Pacific atolls, the extraordinary healing powers of nature have been demonstrated. Islands devastated by radioactivity, heat, and blast and swept clean by water have revealed no evidence that normal growth is not occurring. The probabilities of remote radiation effects could not be denied, but no positive evidence of such effects was found in the test atolls or anywhere else in the Pacific.

One of the basic difficulties in interpreting the Pacific experience unambiguously is that prior to nuclear testing there had been no opportunity to establish natural backgrounds and ecological baselines. In recognition of the need to adequately document the effects, or lack of them, in Plowshare experiments, when the first large cratering experiment was proposed (the Chariot experiment in Alaska) it was decided as a basic part of the experiment to establish all backgrounds before the experiment was conducted. Included in the investigation were those studies also needed to assure that the experiment could be safely conducted. The work was carried out under the auspices of a biological-ecological panel led by John Wolfe of the Division of Biology and Medicine of the AEC. A highly competent and detailed job was done comprising three seasons beginning in the late spring of 1959 and at a cost of several millions of dollars. The conclusion of this group, published early in 1963,[7] was that the experiment as finally proposed "warrants the conclusion that if the detonation were carried out, the chance of biological cost at the ecological level, including jeopardy to the Eskimos or the plants and animals from which they derive their livelihood, appears exceedingly remote. There are necessarily some uncertainties involved in some predictions, however, that can be resolved only by experimentation."

The Chariot experiment has been shelved, not because of the possible biological impact but because it has been overtaken by events. From a technical standpoint, with the exception of the documentation of the explosion effects on the ecology, much of the experimental data have now been obtained or soon will be from experiments in Nevada. An important and perhaps key difficulty in the path of the Plowshare Program was demonstrated by the Chariot experience, namely, the relevance of public information. Fears concerning radioactivity were generated in the minds of the local population which were impossible to alleviate. This experience emphasizes what I believe to be the greatest problem in the Plowshare Program—that of establishing public confidence and acceptance.

This problem can perhaps best be approached

through the recognition that development of explosives with much reduced radioactivity and refined methods of emplacement are expected to reduce fallout in the vicinity of craters to acceptable levels. By acceptable levels is meant that excavation projects might be accomplished without exceeding present tolerances for radiation exposure recommended by the Federal Radiation Council.

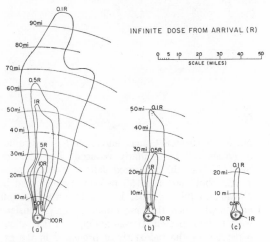

Fig. 4

However, to establish public confidence, care must be taken in any future experiments and demonstrations that all appropriate backgrounds are acquired before the experiments are conducted and that all appropriate bio-ecological work is encouraged. The preparatory work for the Chariot experiment provides a good example of such necessary exhaustive preparation.

To illustrate how advances in explosives technology and mode of emplacement can affect fallout in excavation projects I shall refer to recent results. In Figs. 4 and 5 (courtesy Lawrence Radiation Laboratory) are shown the fallout patterns for SEDAN, which was the first large-scale Plowshare excavation experiment, together with patterns to be expected with advanced explosives. It is noted that advanced nuclear explosives and the technology of emplacement will permit major reduction of fission activity deposited above the ground. Figure 6 (courtesy Lawrence Radiation Laboratory) illustrates the general effect of increasing depth of burial of the charge on fraction of radioactivity

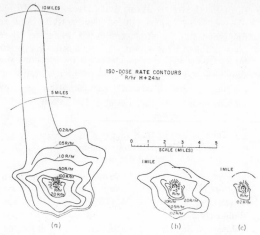

ISO-DOSE RATE CONTOURS
R/hr H+24hr

Fig. 5

out to the surface. Concurrently, the induced radio-activities will be similarly reduced. The plot for SEDAN was the measured fallout pattern for a 100-kt shot (less than 30 percent fission) fired in Nevada on July 6, 1962. This explosion was set off at a depth such that maximum crater dimensions were to be expected. It was anticipated that about 6 percent of the radioactivity produced in the explosion would be released into the cloud formed by the explosion, with the remainder of the radioactivity remaining underground at the site of the explosion. The explosion provided some surprises

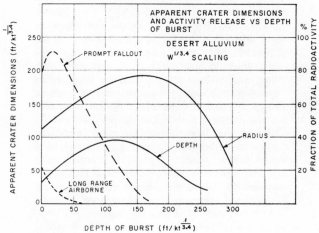

Fig. 6

and therefore new information. The cloud rose considerably higher than expected ($\sim$15 000 feet) because on emergence the gases were hotter than anticipated. The depth of the crater came out about as predicted but the diameter and volume were considerably less. In Table 3 the predicted results are compared witn the observed behavior.

Table 3. The SEDAN shot was 100 kt placed at a depth of 635 feet down a 36-inch diameter drilled and cased hole

Crater Dimensions	Predicted Dimensions	Observed Dimensions
Depth (feet)	310	320
Radius (feet)	735	600
Volume (yards^3)	9 700 000	6 700 000
Cloud Height (feet)	$\sim$4000	$\sim$15 000
Fraction of radioactive fallout out of crater	6%	Slightly more

The crater volume for 100 kt is almost half that expected in 1960 [3] and 70 percent of that predicted just before the event. Considering that the predictions were based on an extrapolation from experience with 1-kt nuclear charges and on chemical-explosive experiments at lower yields, the results are encouraging. The two major surprises, namely, the greater height of the cloud and the smaller diameter of the crater, illustrate the value of experiments at full scale. Deeper burial by moderate amounts would be expected to reduce the height of the cloud significantly without major change in crater dimension, but only additional experience with 100-kt to 1-Mt charges can provide quantitative answers.

One other nuclear event at low yield occurred in the last year. A shot yielding 420 tons in basalt at a depth of 112 feet was fired for military purposes, but the results are also of interest to Plowshare. This was the first nuclear shot in a hard, dry medium and there was some question as to how it would behave, particularly in comparison with chemical explosives. Previous chemical high-explosive tests in the same medium suggested that the explosion would produce a crater 116 feet in radius and 58 feet deep. The observed values for diameter and depth were 110 feet and 63 feet, respectively, indicating that the nuclear explosive behaved very much like the chemical explosive in hard rock. (It is important to note, however, that basic information for chemical explosives in basalt is not well established, so the close agreement may have been fortuitous. The Corps of Engineers is currently carrying out a program with chemical explosives to acquire the needed information.)

While there are differences in the fundamental mechanisms of chemical and nuclear explosives, the differences are well enough understood to permit effective use of chemical explosives to make predictions with respect to yields and placement of nuclear shots to achieve desired crater dimensions. It is noted that crater dimensions for nuclear shots in alluvium and hard rock appear to be the same for the same test conditions, suggesting that medium effects for nuclear explosions may not be very important. The scaled dimensions of nuclear explosions in (alluvium) and in (basalt) agree to better than 10 percent.

To round out the picture technically, large shots, 100 kt to 1 Mt, will be necessary in hard rock, carbonate rocks, and, although of less interest, soft, saturated material. Some of these results could be acquired as part of needed excavation projects and could also serve as demonstration experiments.

The future of excavation by nuclear methods will depend upon developing public confidence in the process through careful public information, backed up by well documented bio-ecological programs and appropriate pragmatic demonstration projects involving large-scale nuclear explosions and a continuing active program of nuclear-explosive development to reduce radioactivities and costs to a minimum.

A critical step in establishing confidence will be the declassification of relevant information, specifically information that will provide the detailed quantities, identification by isotopes, and distribution of radioactivities from advanced explosives. Since much of this information is now considered revealing in terms of the sensitive characteristics of nuclear weapons, it has not been released. Unless ways are found to provide for public scrutiny, debate, and assessment of all of the factors which may bear on the problem of possible radiological effects, I do not believe there is much chance for great progress in the program.

Whether the Plowshare excavation program can in fact be carried out must necessarily depend upon the national policy with respect to nuclear explosions for peaceful purposes. There is little doubt that the present state of the art makes feasible many projects, but, as development, experiment, and demonstration proceed, many additional projects can be seriously considered. In my opinion, this program, adequately supported and actively pursued, could in the near future begin to repay the American people for their investment in nuclear energy for peaceful purposes and could make important contributions to peace and international cooperation.

The author is indebted to the Plowshare staff of the Lawrence Radiation Laboratory, Livermore, California, for providing the basic technical information on nuclear explosives technology on which this paper is based. In addition, special gratitude is expressed to John Foster, Roger Batzel, Gary Higgins, Edward Teller, and Milo Nordyke for reading the manuscript and offering many valuable suggestions, as well as providing several graphs. The ecological information was derived from reports and discussions with Lauren Donaldson and Neal O. Hines of the Laboratory of Radiation Biology of the University of Washington, and John Wolfe of the Division of Biology and Medicine, Atomic Energy Commission.

References

1. J. Herbert Holloman, "Science, Technology, and Economic Growth," Physics Today, March 1963, p. 38.
2. M. D. Nordyke, "An Analysis of Cratering Data from Desert Alluvium," J. Geophys. Res. **67**, 1965 (1963).
3. G. W. Johnson, "Excavation with Nuclear Explosives," University of California Report MCRL-5917, November 1, 1960.
4. Report on a Long-Range Program for Isthmian Canal Transits, House Report No. 1960 of the 86th Congress, 2nd Session, June 23, 1960.
5. Annual Report to Congress for 1962 of the US Atomic Energy Commission.
6. Neal O. Hines, *Proving Ground, An Account of the Radiobiological Studies in the Pacific 1946–1961*. University of Washington Press, Seattle, 1962.
7. Bette Weichold, ed., "Bioenvironmental Features of the Ogotoruk Creek Area, Cape Thompson, Alaska." TID-17226, US Atomic Energy Commission—Division of Technical Information (October, 1962).